Advances in Intelligent Systems and Computing

Volume 1109

The series "Advances in Intelligent Systems and Computing" contains publications on theory, applications, and design methods of Intelligent Systems and Intelligent Computing. Virtually all disciplines such as engineering, natural sciences, computer and information science, ICT, economics, business, e-commerce, environment, healthcare, life science are covered. The list of topics spans all the areas of modern intelligent systems and computing such as: computational intelligence, soft computing including neural networks, fuzzy systems, evolutionary computing and the fusion of these paradigms, social intelligence, ambient intelligence, computational neuroscience, artificial life, virtual worlds and society, cognitive science and systems, Perception and Vision, DNA and immune based systems, self-organizing and adaptive systems, e-Learning and teaching, human-centered and human-centric computing, recommender systems, intelligent control, robotics and mechatronics including human-machine teaming, knowledge-based paradigms, learning paradigms, machine ethics, intelligent data analysis, knowledge management, intelligent agents, intelligent decision making and support, intelligent network security, trust management, interactive entertainment, Web intelligence and multimedia.

The publications within "Advances in Intelligent Systems and Computing" are primarily proceedings of important conferences, symposia and congresses. They cover significant recent developments in the field, both of a foundational and applicable character. An important characteristic feature of the series is the short publication time and world-wide distribution. This permits a rapid and broad dissemination of research results.

** Indexing: The books of this series are submitted to ISI Proceedings, EI-Compendex, DBLP, SCOPUS, Google Scholar and Springerlink **

More information about this series at http://www.springer.com/series/11156

Siddhartha Bhattacharyya · Sushmita Mitra ·
Paramartha Dutta
Editors

Intelligence Enabled Research

DoSIER 2019

 Springer

Editors
Siddhartha Bhattacharyya
RCC Institute of Information Technology
Kolkata, India

Sushmita Mitra
Indian Statistical Institute
Kolkata, India

Paramartha Dutta
Visva-Bharati University
Santiniketan, India

ISSN 2194-5357   ISSN 2194-5365 (electronic)
Advances in Intelligent Systems and Computing
ISBN 978-981-15-2020-4   ISBN 978-981-15-2021-1 (eBook)
https://doi.org/10.1007/978-981-15-2021-1

This Springer imprint is published by the registered company Springer Nature Singapore Pte Ltd.
The registered company address is: 152 Beach Road, #21-01/04 Gateway East, Singapore 189721, Singapore

Siddhartha Bhattacharyya would like to dedicate this book to his father Late Ajit Kumar Bhattacharyya, his mother Late Hashi Bhattacharyya, and his beloved wife Rashni Bhattacharyya

Sushmita Mitra would like to dedicate this book to her father Late Dr. Girindra Nath Mitra and her mother Late Dr. Maya Mitra

Paramartha Dutta would like to dedicate this book to his father Late Arun Kanti Dutta and his mother Late Bandana Dutta

Organization Committees

Chief Patron

Dr. Saikat Maitra, Vice Chancellor, MAKAUT, Kolkata, India
Dr. Ajoy K. Ray, RCC Institute of Information Technology, Kolkata, India

Honorary Chair

Dr. Vincenzo Piuri, Università degli Studi di Milano, Milano, Italy

General Chairs

Dr. Siddhartha Bhattacharyya, RCC Institute of Information Technology, Kolkata, India
Dr. Sushmita Mitra, ISI Kolkata, India
Dr. Paramartha Dutta, Visva-Bharati University, Santiniketan, India

Program Chairs

Dr. Alok Kole, RCC Institute of Information Technology, Kolkata, India
Dr. Indrajit Pan, RCC Institute of Information Technology, Kolkata, India
Dr. Abhishek Basu, RCC Institute of Information Technology, Kolkata, India
Dr. Srijan Bhattacharya, RCC Institute of Information Technology, Kolkata, India

Technical Co-chairs

Dr. Ashoke Mondal, RCC Institute of Information Technology, Kolkata, India
Dr. Soham Sarkar, RCC Institute of Information Technology, Kolkata, India
Dr. Shilpi Bhattacharya Deb, RCC Institute of Information Technology, Kolkata, India
Dr. Minakshi Banerjee, RCC Institute of Information Technology, Kolkata, India
Dr. Dipankar Majumdar, RCC Institute of Information Technology, Kolkata, India

Organizing Secretaries

Dr. Srijibendu Bagchi, RCC Institute of Information Technology, Kolkata, India
Dr. Debasish Mondal, RCC Institute of Information Technology, Kolkata, India
Dr. Abhijit Das, RCC Institute of Information Technology, Kolkata, India
Prof. Rajib Saha, RCC Institute of Information Technology, Kolkata, India
Prof. Arijit Ghosh, RCC Institute of Information Technology, Kolkata, India
Prof. Arup Bhattacharjee, RCC Institute of Information Technology, Kolkata, India

International Advisory Committee

Dr. Cesare Alippi, Politecnico di Milano, Italy
Dr. Debotosh Bhattacharjee, Jadavpur University, India
Dr. Aboul Ella Hassanien, Cairo University, Egypt
Dr. Saman Halgamuge, The University of Melbourne, Australia
Dr. Elizabeth Behrman, Wichita State University, USA
Dr. Mario Koeppen, Kyushu Institute of Technology, Japan

Technical Program Committee

Dr. Nilanjan Dey, Techno International, Kolkata, India
Dr. Sourav De, Cooch Behar Government Engineering College, India
Dr. Jyoti Prakash Singh, NIT Patna, India
Dr. Rik Das, Xavier Institute of Social Sciences, India
Dr. Anasua Sarkar, Jadavpur University, India
Dr. Eduard Babulak, Liberty University, USA
Dr. Xiao-Zhi Gao, University of Eastern Finland, Finland
Dr. Koushik Mondal, IIT Dhanbad, India
Dr. Kousik Dasgupta, Kalyani Government Engineering College, India

Dr. Leo Mrsic, Algebra University College, Croatia
Dr. Debashis De, MAKAUT, India
Dr. Mihaela Albu, Politehnica University of Bucharest, Romania
Dr. Mariofanna Milanova, University of Arkansas at Little Rock, USA
Dr. Ashish Mani, Amity University, India
Dr. Anupam Karmakar, Calcutta University, India
Dr. C. T. Singh, Sikkim Manipal Institute of Technology, East Sikkim, India
Dr. Ratika Pradhan, Sikkim Manipal Institute of Technology, East Sikkim, India
Dr. Somenath Chatterjee, Sikkim Manipal Institute of Technology,
East Sikkim, India
Dr. A. K. Roy, Sikkim Manipal Institute of Technology, East Sikkim, India
Dr. Rabindranath Bera, Sikkim Manipal Institute of Technology, East Sikkim, India
Dr. Karma Sonam Sherpa, Sikkim Manipal Institute of Technology,
East Sikkim, India
Dr. Bibhu Prasad Swain, Sikkim Manipal Institute of Technology,
East Sikkim, India
Dr. Nitai Paitya, Sikkim Manipal Institute of Technology, East Sikkim, India
Dr. Arindam Mondal, RCC Institute of Information Technology, India
Dr. Ruhul Islam, Sikkim Manipal Institute of Technology, East Sikkim, India
Dr. Anirban Mukherjee, RCC Institute of Information Technology, Kolkata, India
Dr. Dipankar Santra, RCC Institute of Information Technology, Kolkata, India
Dr. Anup Kolya, RCC Institute of Information Technology, Kolkata, India
Dr. Shashank Pushkar, Birla Institute of Technology, Mesra, India
Dr. Goran Klepac, Algebra University College, Croatia
Dr. Jan Platos, VSB Technical University of Ostrava, Czech Republic
Dr. Samerjeet Borah, Sikkim Manipal Institute of Technology, East Sikkim, India
Dr. Mousumi Gupta, Sikkim Manipal Institute of Technology, East Sikkim, India
Dr. Mohan Pratap Pradhan, Sikkim University, India
Dr. Sandip Dey Global, Sukanta Mahavidyalaya, Jalpaiguri, India
Mr. Arpan Deyasi, RCC Institute of Information Technology, Kolkata, India
Mr. Debanjan Konar, Sikkim Manipal Institute of Technology, East Sikkim, India
Mr. Udit Kr. Chakraborty, Sikkim Manipal Institute of Technology,
East Sikkim, India
Mr. Hirak Dasgupta, Indian Institute of Technology Bombay, India
Mr. Teerthankar Ghosal, Indian Institute of Technology Patna, India
Dr. Rajarshi Mahapatra, IIIT Naya Raipur, India

Publicity and Sponsorship Chairs

Dr. Anup Koyla, RCC Institute of Information Technology, Kolkata, India
Mr. Soumen Mukherjee, RCC Institute of Information Technology, Kolkata, India
Mr. Manas Ghosh, RCC Institute of Information Technology, Kolkata, India
Mr. Sujoy Mondal, RCC Institute of Information Technology, Kolkata, India

Local Hospitality Chairs

Mr. Pankaj Pal, RCC Institute of Information Technology, Kolkata, India
Ms. Alokananda Dey, RCC Institute of Information Technology, Kolkata, India
Mr. Amit Khan, RCC Institute of Information Technology, Kolkata, India
Mr. Falguni Adhikary, RCC Institute of Information Technology, Kolkata, India
Mr. Deepam Ganguly, RCC Institute of Information Technology, Kolkata, India

Finance Chair

Mr. Chinmoy Ghosal, RCC Institute of Information Technology, Kolkata, India

Organizing Committee

Mr. Arindam Mondal, RCC Institute of Information Technology, Kolkata, India
Mr. Soumen Mukherjee, RCC Institute of Information Technology, Kolkata, India
Mr. Arup Kr Bhattacharjee, RCC Institute of Information Technology,
Kolkata, India
Mr. Jayanta Datta, RCC Institute of Information Technology, Kolkata, India
Mr. Ranjan Jana, RCC Institute of Information Technology, Kolkata, India
Mr. Biswanath Chakraborty, RCC Institute of Information Technology,
Kolkata, India
Mrs. Alokananda Dey, RCC Institute of Information Technology, Kolkata, India
Mrs. Satabdwi Sarkar, RCC Institute of Information Technology, Kolkata, India
Mrs. Arpita Banerjee, RCC Institute of Information Technology, Kolkata, India
Mrs. Pampa Debnath, RCC Institute of Information Technology, Kolkata, India
Mrs. Saraswati Saha, RCC Institute of Information Technology, Kolkata, India
Dr. Arpita Ghosh, RCC Institute of Information Technology, Kolkata, India
Dr. Tiya Dey Malakar, RCC Institute of Information Technology, Kolkata, India
Mr. Nandan Bhattacharyya, RCC Institute of Information Technology,
Kolkata, India
Ms. Jayanti Das, RCC Institute of Information Technology, Kolkata, India
Mr. Anindya Basu, RCC Institute of Information Technology, Kolkata, India
Mr. Harinandan Tunga, RCC Institute of Information Technology, Kolkata, India
Dr. Pramit Ghosh, RCC Institute of Information Technology, Kolkata, India
Mrs. Monika Singh, RCC Institute of Information Technology, Kolkata, India
Ms. Parama Bagchi, RCC Institute of Information Technology, Kolkata, India
Mr. Somenath Nag Choudhury, RCC Institute of Information Technology,
Kolkata, India

Mrs. Abantika Choudhury, RCC Institute of Information Technology,
Kolkata, India
Ms. Moumita Deb, RCC Institute of Information Technology, Kolkata, India
Mr. Hiranmoy Roy, RCC Institute of Information Technology, Kolkata, India
Mr. Soumyadip Dhar, RCC Institute of Information Technology, Kolkata, India
Mr. Sudarsan Biswas, RCC Institute of Information Technology, Kolkata, India
Mr. Nijam Ud-Din Molla, RCC Institute of Information Technology, Kolkata, India
Mr. Budhaditya Biswas, RCC Institute of Information Technology, Kolkata, India
Mr. Sarbojit Mukherjee, RCC Institute of Information Technology, Kolkata, India
Mr. Subhasis Bandopadhyay, RCC Institute of Information Technology,
Kolkata, India
Mr. Debabrata Bhattacharya, RCC Institute of Information Technology,
Kolkata, India
Mr. Kalyan Biswas, RCC Institute of Information Technology, Kolkata, India
Ms. Naiwrita Dey, RCC Institute of Information Technology, Kolkata, India
Mr. Avishek Paul, RCC Institute of Information Technology, Kolkata, India
Dr. Tathagata Deb, RCC Institute of Information Technology, Kolkata, India
Dr. Kanchan Kumar Patra, RCC Institute of Information Technology,
Kolkata, India
Dr. Papia Datta, RCC Institute of Information Technology, Kolkata, India
Ms. Satarupa Chatterjee, RCC Institute of Information Technology, Kolkata, India
Dr. Sangita Agarwal, RCC Institute of Information Technology, Kolkata, India
Dr. Anirban Mukherjee, RCC Institute of Information Technology, Kolkata, India
Ms. Jhuma Ray, RCC Institute of Information Technology, Kolkata, India
Dr. Sadhan Kumar Dey, RCC Institute of Information Technology, Kolkata, India
Mr. Avijit Saha, RCC Institute of Information Technology, Kolkata, India
Mr. Nitai Banerjee, RCC Institute of Information Technology, Kolkata, India
Ms. Anwesha Basu, RCC Institute of Information Technology, Kolkata, India
Dr. Sayantani Maity, RCC Institute of Information Technology, Kolkata, India
Ms. Shaswati Roy, RCC Institute of Information Technology, Kolkata, India
Mr. Souvik Majumdar, RCC Institute of Information Technology, Kolkata, India

Preface

With the advent and development of computational intelligence, almost every technological innovation in the present times is being driven by intelligence in one form or the other. Of late, computational intelligence has made its presence felt in every nook and corner of the world, thanks to the rapid exploration of research in this direction. Computational intelligence is now not limited to only specific computational fields, it has made inroads in signal processing, smart manufacturing, predictive control, robot navigation, smart cities, sensor design, to name a few. Keeping in mind the stress laid out by the Government of India on promotion and use of computational intelligence, several efforts are being invested for the same. The proposed Doctoral Symposium is one such attempt in this direction.

The 2019 First Doctoral Symposium on Intelligence Enabled Research (DoSIER 2019) was organized by RCC Institute of Information Technology, Kolkata, India, during October 19–20, 2019. The symposium aimed to provide doctoral students and early career researchers an opportunity to interact with their colleagues working on foundations, techniques, tools, and applications of computational intelligence.

The goals of the symposium centered on:

1. Providing the participants independent and constructive feedback on their current research and future research directions.
2. Developing a supportive community of scholars and a spirit of collaborative research.
3. Providing an opportunity for student participants to interact with established researchers and practitioners in relevant fields of research.
4. Opening up the possibility of a closed research forum for mutual exchange of knowledge base and know-hows.

The symposium was technically sponsored by IEEE Computational Intelligence, Kolkata Chapter.

DoSIER 2019 has observed eight keynote sessions by eminent researchers and academicians across the globe along with two technical tracks and one panel discussion session with eminent academicians. The keynote speakers included (i) Prof. Ernesto Cuadros Verges, University of Engineering and Technology, Lima,

Peru; (ii) Dr. Sushmita Mitra, FIEEE, Indian Statistical Institute, Kolkata, India; (iii) Dr. Prabodh Bajpai, IIT Kharagpur, India; (iv) Dr. Ashish Mani, Amity University, Noida, India; (v) Dr. Vincenzo Piuri, FIEEE, Università degli Studi di Milano, Milano, Italy; (vi) Dr. Atiqur Rahman Ahad, University of Dhaka, Bangladesh; and (vii) Mr. Sabyasachi Mukhopadhyay, BIMS, Kolkata.

DoSIER 2019 received a good number of submissions from the doctoral students in the country. After peer review, only 14 papers were accepted to be presented in the conference. Authors from different parts of the country presented their peer-reviewed articles under two sessions of DoSIER 2019.

A highly informative panel discussion session was held on the second day of the event. Prof. Ujjwal Maulik and Prof. Debotosh Bhattacharjee, both from Jadavpur University, India, were present as panel experts. The participants enjoyed and gained knowledge and understanding for achieving quality research outcomes.

Three best papers were awarded in 2019 First Doctoral Symposium on Intelligence Enabled Research (DoSIER 2019). These awards were sponsored by IFERP, India.

Kolkata, India	Siddhartha Bhattacharyya
Kolkata, India	Sushmita Mitra
Santiniketan, India	Paramartha Dutta
October 2019	

Contents

About the Editors

Dr. Siddhartha Bhattacharyya is currently the Principal of RCC Institute of Information Technology, Kolkata, India. He served as a Senior Research Scientist at the Faculty of Electrical Engineering and Computer Science of VSB Technical University of Ostrava, Czech Republic, from October 2018 to April 2019. Prior to this, he was the Professor of Information Technology at RCC Institute of Information Technology, Kolkata, India. He is a co-author of 5 books and co-editor of 30 books, and has more than 220 research publications in international journals and conference proceedings to his credit. His research interests include soft computing, pattern recognition, multimedia data processing, hybrid intelligence, and quantum computing.

Dr. Sushmita Mitra is the Head of and INAE Chair Professor at the Machine Intelligence Unit (MIU), Indian Statistical Institute, Kolkata. From 1992 to 1994, she was a DAAD Fellow at the RWTH in Aachen, Germany. She was a Visiting Professor at the Computer Science Department of the University of Alberta, Edmonton, Canada, in 2004 and 2007; Meiji University, Japan, in 1999, 2004, 2005 and 2007; and Aalborg University Esbjerg, Denmark, in 2002 and 2003. She has more than 150 research publications in refereed international journals to her credit.

According to the Science Citation Index (SCI), two of her papers have been ranked 3rd and 15th in the list of top-cited papers in Engineering Science from India during 1992–2001. Dr. Mitra is a Fellow of the IEEE, Indian National Science Academy (INSA), International Association for Pattern Recognition (IAPR), Indian National Academy of Engineering (INAE), and National Academy of Sciences, India (NASI). She was an IEEE CIS Distinguished Lecturer for the period 2014–2016. Dr. Mitra is the current Chair of the IEEE CIS Kolkata Chapter. Her current research interests include data mining, pattern recognition, soft computing, medical image processing, and bioinformatics.

Dr. Paramartha Dutta received his bachelor's and master's degrees in Statistics from the Indian Statistical Institute, Calcutta, in 1988 and 1990, respectively. He completed his Master of Technology in Computer Science at the same institute in

1993, and his Doctor of Philosophy in Engineering at Bengal Engineering and Science University, Shibpur, in 2005. He has served on various projects funded by the Government of India, e.g. for the Defence Research and Development Organization, Council of Scientific and Industrial Research, Indian Statistical Institute, Calcutta, etc. Dr. Dutta is currently a Professor at the Department of Computer and System Sciences, Visva Bharati University, West Bengal, India. He has co-authored four books and has one edited book to his credit. He has published ca. 100 papers in various journals and conference proceedings, both national and international.

Vehicle Pollution Detection from Images Using Deep Learning

Srimanta Kundu and Ujjwal Maulik

Abstract Vehicle pollution is one of the biggest contributors among the Air pollution sources. The main objective of this study is, to identify the pollutant vehicle from on-road real time images. We propose a novel image-based transfer learning approach by identifying the emission from the vehicle. These images can be captured from other nearby or adjacent vehicles or from traffic control units. Once the pollutant vehicle is detected, this information can be used for notification, pollution control, and surveillance in future as well. Our deep learning-based method involves Inception-v3, and it can work under any weather and light conditions with varying environments.

Keywords Inception-v3 · Transfer learning · Vehicle pollution · Deep learning · Saliency map

1 Introduction

One of the major concerns of Intelligent Transportation Systems (ITS) is to reduce the pollution emitted from the vehicle. Despite the tremendous progress in the field of pollution detection, the area remains elusive. Currently, researchers from different corners have put their effort to detect pollution sources and quality using various means. Zhang et al. [1] have provided one convolution network-based model to estimate air quality. Kök et al. [2] presented a deep learning mechanism based on IoT data to predict air quality. Li et al. have presented a comparison with the spatiotemporal artificial neural network (STANN) of air quality prediction [3]. Kalapanidas [4] proposed short term NO_2 concentration prediction and a case-based classifier approach to measure the air feature.

S. Kundu (✉) · U. Maulik
Jadavpur University, Kolkata 700032, India
e-mail: srimantacse@yahoo.co.in

U. Maulik
e-mail: umaulik@cse.jdvu.ac.in

© Springer Nature Singapore Pte Ltd. 2020
S. Bhattacharyya et al. (eds.), *Intelligence Enabled Research*,
Advances in Intelligent Systems and Computing 1109,
https://doi.org/10.1007/978-981-15-2021-1_1

Here, we have proposed an image-based deep learning approach to detect the pollution that is generated from on-road vehicles. We have used the transfer learning-based Long Short Term Memory (LSTM) to generate the prediction model. The rest of the paper is structured as follows. Section 2 contains the methodology. In Sect. 3, we have shared experimental setup and database processing. We have presented the result of multiple runs in Sect. 4. Finally, Sect. 5 concludes the paper.

2 Methodology

In the proposed approach, initially, the Inception-v3 model has been used, which is a well-known image recognition model provided by Google. The most important feature of Inception-v3 is factorization into smaller convolutions. It emerges in the research of Szegedy et al. [5], develops over the years and becomes inevitable with greater than 78.1% accuracy on the Image-Net dataset [6–10]. Transfer learning is a popular machine learning approach where developed models can contribute as the initial weights and enhance the output of other models [11, 12]. Some design changes may be included in the architecture itself for better efficiency as per the domain expertise. The concept of transfer learning is used in the proposed model. We have chosen Inception-v3 as the initial weight values and did the required adjustment in the dense and activation layers after using LSTM. Finally, the model is trained with a variety of training data until it got stabilized.

3 Database and Experimental Setup

The proposed model is trained with our self-made database, mostly collected from the internet. In this regard, different types of emissions from different vehicles are used. Besides these, few videos are collected in real time during the journey through the new town road, Kolkata. In those real time data, we have tried to consider different environmental conditions like cloudy, rainy, day, night, etc. Some sample images have been shown in Fig. 1. For this experiment, we have used total 200 images [90 images from Class 1, i.e., No Pollution from Vehicle and 110 images from Class 2, i.e., Pollution from Vehicle Only]. We have shown the results, one set with 70% training data and another with 80% training data.

While using transfer learning, we have used another connected network at the end of the Inception-v3 layers. The last layer of Inception-v3 has been peeled off and mapped to two nodes which denotes the two output classes. The basic mode of the architecture is LSTM followed by three sets of dense and activation layers with different number of nodes. The detailed shape has been provided in Table 1.

The algorithm runs 1500 iterations to minimize the Mean Absolute Error (MAE).

(a)   **(b)**

Fig. 1 **a** Class 1: no pollution from the car with varying environments. **b** Class 2: pollution from the car only

Table 1 Proposed deep learning network architecture

Layer	LSTM_1	Densel_1	Activation_1	Dense_2	Activation_2	Dense_3	Activation_3
Output shape	(None, 512)	(None, 1024)	(None, 1024)	(None, 50)	(None, 50)	(None, 2)	None, 2)
Parameters	5,244,928	525,312	0	51,250	0	102	0

4 Result

The learning capability of deep neural network has been demonstrated using a well-known static saliency map [13] concept. From Fig. 2 we can see three images in each row, while the first one (Ai, $i = 1, 2$) is the original image, the second image (Bi) is

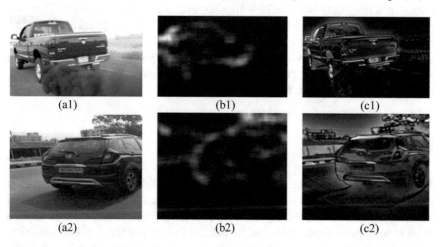

(a1)   (b1)   (c1)

(a2)   (b2)   (c2)

Fig. 2 Visualization of the training model: **a**1 original image, **b**1 static saliency map, **c**1 saliency map after image threshholding

Table 2 Results of two runs

Training Percentage	Testing Accuracy	Precision	Recall	F Score	Confusion Matrix	
Run 1: 80%	0.975	1.0	0.952	0.952	20	0
					1	19
Run 2: 70%	0.933	1.0	0.882	0.882	30	0
					4	26

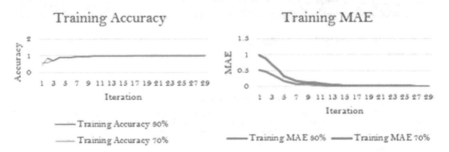

Fig. 3 Accuracy and MAE trend with no of iterations

the static saliency map learned by the CNN model. Subsequently, the third one (Ci) is generated using image thresholding technique on top of the second image.

During the experiment, we have run the Deep learning model several times with different ratio of splitting the training and testing data set. Two sample runs' result has been shown. With 80% training sample we have achieved 97% accuracy in the unknown testing data set. For 70% training data the accuracy is 93%. From the confusion matrix, as shown in Table 2, in each run we have computed the Precision, Recall, and F-Score [14].

The graph in Fig. 3 shows how the model gradually gains the maximum accuracy with number of iterations during the training period. At the same time, the MAE values stabilize to a certain point.

5 Conclusion

The paper demonstrates the utility of Deep Learning for the pollution detection of vehicles from real time images. The result clearly shows that simple deep neural network architecture has the capability of providing significant results. In future a more sophisticated architecture can be defined to improve the performance. Morphological and structural analysis is another part where the focus can be put on the smoke and the location of the same to properly identify as vehicle pollution.

References

1. C. Zhang, J. Yan, X. Rui, L. Liu, R. Bie, On estimating air pollution from photos using convolutional neural network, in *Proceedings of the 24th ACM International Conference on Multimedia*, ACM (2016), pp. 297–301
2. İ. Kök, M.U. Şimşek, S. Özdemir, A deep learning model for air quality prediction in smart cities, in *IEEE International Conference on Big Data (Big Data)*, IEEE (2017), pp. 1983–1990
3. X. Li, L. Peng, Y. Hu, J. Shao, T. Chi, Deep learning architecture for air quality predictions. Environ. Sci. Pollut. Res. **23**(22), 22408–22417 (2016)
4. E. Kalapanidas, N. Avouris, Short-term air quality prediction using a case-based classifier. Environ. Model Softw. **16**(3), 263–272
5. C. Szegedy, V. Vanhoucke, S. Ioffe, J. Shlens, Z. Wojna, Rethinking the inception architecture for computer vision, in *Proceedings of the IEEE Conference on Computer Vision and Pattern Recognition* (2016), pp. 2818–2826
6. A. Krizhevsky, I. Sutskever, G.E. Hinton, Imagenet classification with deep convolutional neural networks, in *Advances in Neural Information Processing Systems* (2012), pp. 1097–1105
7. K. Simonyan, A. Zisserman, *Very deep convolutional networks for large-scale image recognition.* arXiv:1409.1556 (2014)
8. C. Szegedy, W. Liu, Y. Jia, P. Sermanet, S. Reed, D. Anguelov, D. Erhan, V. Vanhoucke, A. Rabinovich, Going deeper with convolutions, in *Proceedings of the IEEE Conference on Computer Vision and Pattern Recognition* (2015), pp. 1–9
9. S. Ioffe, C. Szegedy, *Batch normalization: accelerating deep network training by reducing internal covariate shift.* arXiv:1502.03167 (2015)
10. K. He, X. Zhang, S. Ren, J. Sun, Deep residual learning for image recognition, in *Proceedings of the IEEE Conference on Computer Vision and Pattern Recognition* (2016), pp. 770–778
11. S.J. Pan, Q. Yang, A survey on transfer learning. IEEE Trans. Knowl. Data Eng. **22**(10), 345–1359 (2009)
12. C.D. Gürkaynak, N. Arica, A case study on transfer learning in convolutional neural networks, in *26th Signal Processing and Communications Applications Conference (SIU)*, IEEE (2018), pp. 1–4
13. T.V. Nguyen, M. Xu, G. Gao, M. Kankanhalli, Q. Tian, S. Yan, Static saliency vs dynamic saliency: a comparative study, in *Proceedings of the 21st ACM International Conference on Multimedia*, ACM (2013), pp. 987–996
14. C. Goutte, E. Gaussier, A probabilistic interpretation of precision, recall and F-score, with implication for evaluation, in *European Conference on Information Retrieval* (Springer, Heidelberg, 2005), pp. 345–359

Single Image De-raining Using GAN for Accurate Video Surveillance

Ratnadeep Dey and Debotosh Bhattacharjee

Abstract Video surveillance is a very crucial area of research. The accuracy is the basic criterion of this type of research works because the video surveillance system ensures security issues. A less accurate surveillance system may lead to mass causalities in recent terrorism threatening the world. However, the surveillance cameras are installed mainly in open areas. Therefore, the acquired data can be suffered from environmental stimuli. The degraded data lowers the performance of the surveillance system. It is needed to restore the surveillance data for accurate analysis. In this paper, we concentrate on video surveillance data affected by rain. In this work, we de-rain degraded rainy surveillance video data using a Generative Adversarial Network (GAN) model. The proposed model performs better than other types of CNN based de-raining algorithm.

Keywords Video · De-raining · GAN · Video surveillance · Video denoising

1 Introduction

According to physical properties and effects [1], two types of weather conditions can degrade the video surveillance data. The two types are—(i) Steady and (ii) Dynamic. The categorization mainly depends on the size of particles present in the air. The presence of unwanted particles in the air may lead to lower visibility. If the size of the particles present in the air is between 1–10 μm. The fog, mist, the haze comes under the category of the Steady weather condition. In the case of the Dynamic weather conditions, the size of the particles present in the air is greater than 10 μm. Rain, Snow, etc. come under this category.

R. Dey (✉) · D. Bhattacharjee
Department of Computer Science and Engineering, Jadavpur University, Kolkata, India
e-mail: ratnadipdey@gmail.com

D. Bhattacharjee
e-mail: debotoshb@hotmail.com

© Springer Nature Singapore Pte Ltd. 2020
S. Bhattacharyya et al. (eds.), *Intelligence Enabled Research*,
Advances in Intelligent Systems and Computing 1109,
https://doi.org/10.1007/978-981-15-2021-1_2

The restoration of the images which have been degraded by the Dynamic weather conditions is more complicated than the restoration process of the images degraded by the Steady condition. In the case of the Dynamic conditions, the size of particles is bigger than the steady conditions. Therefore, the particles are visible; those have degraded the quality of the images. The particles also work as a convex lens that affects the overall appearance of the image.

Under dynamic conditions, rain-affected video surveillance data is the main concern of the present work. In this work, we have used a Generative Adversarial Network (GAN) [2] to de-rain surveillance video data. The GAN has been used to de-rain single images. We have de-rained video surveillance data, which is a very new work in this field. We have been inspired by the work of ID-CGAN [3] and have created a very similar model to de-rain video surveillance data. We have evaluated the model's performance with our own created dataset. The model performs well in comparison to other CNN based de-raining algorithms.

The paper is organized as follows: The next section surveys some related works; Sect. 3 explains the details of the proposed model; Sect. 4 shows results and discussions and finally, Sect. 5 concludes the paper.

2 Related Work

In recent years, convolution neural network-based single image rain removal becomes one of the interesting research topics in computer vision and pattern recognition field. Many researchers contribute in their own way. Fu et al. [4] proposed De-rainNet to remove rain streaks from rainy images. The model is trained with both a rainy image and a clean image and extracts features by comparing both images. The authors evaluated their work on synthetic data. The model fails to de-rain our surveillance video data. Wang et al. [5] de-rained images in multi-steps—courser to finer approach. They used densely connected dilation convolution block to extract detail features of raining images. The authors evaluate their work on public datasets and find good results. Fu et al. [6], in their work, proposes a de-raining model inspired by the res-net model. The authors trained the network with the high frequency details of rainy image. The modification helps to model the rain streaks separated from the background. The recurrent neural network is fused with convolution neural network for single image denoising [7]. The de-raining model can estimate rain streaks in different directions. In work [8], a single image de-raining model named DID-MDN is proposed. The model can estimate the density of the rain. According to the rain density, the model removes rain streaks. The model uses a multi-scale dense network as its backbone. Zeng et al. [3] used GAN based approach for image denoising. They used CGAN network as its basic architecture. This model helps us to build our model.

3 Methodology

I. Goodfellow created a new orientation of research in computer vision and pattern recognition area by proposing the Generative Adversarial Network [2]. The GANmainly designed to generate data from random noise. The GAN consists of two networks—the generator and the discriminator. The discriminator tries to understand the data distribution of real image while the Generator tries to generate data similar to real data from random noise. The discriminator network evaluates the performance of the generator network and guides it to generate data similar to the real one.

3.1 The Basic Idea Behind the Work

Although the GAN is developed to generate new data, it can be used in the image restoration problem. In our work, we trained the discriminator model with video data of normal weather. The discriminator model learns the data distribution from the real data. Then the rainy video frames are inserted in the generator model. The generator model tries to generate real alike data. The discriminator model is trained first followed by the generator model. The generator model produces rain-free data after a couple of iterations. Figure 1 depicts the scenario of the working principle of the model.

3.2 Network Architecture

The network architecture is very similar to the work proposed in [3]. The basic network used here is CGAN. The generator model consists of densely connected

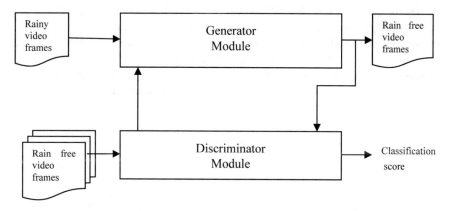

Fig. 1 Basic working principle of the method

(a) Input video frame   (b) Output video frame

Fig. 2 Input and corresponding output video frame

Table 1 Comparison with state-of-the-art techniques

	Proposed method	De-rainNet [4]	DID-MDN [8]
PSNR	**23.93**	22.09	21.28
SSIM	**0.8152**	0.7856	0.7265

symmetric convolution networks. The discriminator model used here is a multi-scale network. Refined perceptual loss is used to stable the training procedure for GAN.

4   Result and Discussion

We have used video data captured by a surveillance camera. Frames are extracted from the video data. We have used 527 rainy video frames and 541 rain-free video frames for training. Near about 250 video frames each for rainy and rain-free video frames that are used for testing. In Fig. 2a, a rainy video frame and in Fig. 2b corresponding de-rained output is shown. The output of the model is compared with the other state-of-the-art CNN based de-raining techniques. Table 1 shows the comparison results. The comparison result ensures that the proposed model outperforms other de-raining techniques in case of this video surveillance data.

5 Conclusion

In this work, we used the GAN for de-raining video surveillance data. Although the GAN is designed for generating new data, it performs well in case of video de-raining. The proposed method proves itself better than the other CNN based networks. The model has some loopholes. We will try to resolve the shortcomings in the near future.

Acknowledgements The authors are thankful to DRDO, the funding authority of the project named "Development of object detection techniques from degraded complex video sequences due to dynamic variation of scenes by different atmospheric conditions for security &surveillance" and Tripura University for providing the database.

References

1. K. Garg, S.K. Nayar, Vision and rain. Int. J. Comput. Vis. Springer (2007). https://doi.org/10.1007/s11263-006-0028-6
2. I. Goodfellow, J. Pouget-Abadie, M. Mirza, B. Xu, D. Warde-Farley, S. Ozair, A. Courville, Y. Bengio, Generative adversarial nets, in *Proceedings of Advances in Neural Information Processing Systems* (2014), pp. 2672–2680. http://papers.nips.cc/paper/5423-generative-adversarial-nets.pdf
3. H. Zhang, V. Sindagi, V.M. Patel, Image de-raining using a conditional generative adversarial network (2017). https://arxiv.org/abs/1701.05957
4. X. Fu, J. Huang, X. Ding, Y. Liao, J. Paisley, Clearing the skies: a deep network architecture for single-image rain removal. IEEE Trans. Image Process. **26**(6) (2017)
5. C. Wang, M. Zhang, Z. Su, G. Yao, Y. Wang, X. Sun, X. Luo, From coarse to fine: a stage-wise deraining net. IEEE Access **7**. https://doi.org/10.1109/access.2019.2922549
6. X. Fu, J. Huang, D. Zeng, Y. Huang, X. Ding, J. Paisley, Removing rain from single images via a deep detail network, in *IEEE Conference on Computer Vision and Pattern Recognition (CVPR)* (2017). https://doi.org/10.1109/cvpr.2017.186
7. X. Li, J. Wu, Z. Lin, H. Liu, H. Zha, Recurrent squeeze-and-excitation context aggregation net for single image deraining. arXiv:1807.05698
8. H. Zhang, V.M. Patel, Density-aware single image de-raining using a multi-stream dense network. arXiv:1802.07412 [cs.CV]

Towards a Framework for Performance Testing of Metaheuristics

Ashish Mani, Nija Mani and Siddhartha Bhattacharyya

Abstract Metaheuristics have been very successful in solving difficult optimization problems by employing randomness in their search mechanism. They find near-optimal solutions in reasonable amount of computation time and resources. Testing of metaheuristics is performed by computing descriptive statistics as well as by comparative testing with other State-of-the-art methods. However, performance guarantees for Metaheuristics are not provided, which limits its commercial potential. An attempt has been made in this paper to propose a framework for Performance Testing of Metaheuristics.

Keywords Evolutionary algorithms · Swarm algorithms · Optimizations · Heuristics

1 Introduction

Optimization problems are routinely solved in most of the human endeavors and its importance is increasing as current civilization faces stiff competition for scarce resources. During the last century, a lot of work has been done to formalize the study of optimization methods [1]. Currently, there are several methods for solving different kinds of optimization problems [2]. However, it has been observed that most optimization problems in real-world are in NP-Hard complexity class for brute force methods like exhaustive search [3] and often the search space is huge, which renders techniques based on enumerative strategies practically useless [4]. Further, these

A. Mani (✉)
Amity University Uttar Pradesh, Noida, India
e-mail: amani@amity.edu

N. Mani
Dayalbagh Educational Institute, Agra, India
e-mail: mani.nija@gmail.com

S. Bhattacharyya
RCC Institute of Information Technology, Kolkata, India
e-mail: dr.siddhartha.bhattacharyya@gmail.com

© Springer Nature Singapore Pte Ltd. 2020
S. Bhattacharyya et al. (eds.), *Intelligence Enabled Research*,
Advances in Intelligent Systems and Computing 1109,
https://doi.org/10.1007/978-981-15-2021-1_3

13

problems are often ill-behaved so gradient-based techniques are often not applicable or they are not guaranteed to produce global optimal solutions [5].

Metaheuristics have proved very useful in solving difficult optimization problems where search space is large and near-optimal solutions are acceptable in last three decades [6]. Metaheuristics solve optimization problems through search by using multiple agents, some randomness in operators and iterations [7]. In contrast to exact methods, which find optimal solutions using a large amount of computational resources, metaheuristics find near-optimal solutions using reasonable computational resources and time and have been very successful. However, metaheuristics do not have any performance guarantee, i.e., there are neither analytical proofs available for their run-time/space complexity nor any probabilistic guarantee is provided [8].

Metaheuristic often uses some randomness in their search process so they are tested by independently running them N times, where N is mostly, 30, 50 or 100. Further, descriptive statistic like Best, Worst, Mean, Median, Success Run, and Standard Deviation is tabulated and often used for testing purposes and also for comparisons [9]. A relatively recent trend is to perform comparative study using inferential statistics [10], i.e., comparison is made between multiple metaheuristics and an attempt is made to show that one of the metaheuristics outperforms other metaheuristics. However, the descriptive and inferential statistics does not give any performance guarantee like what is the surety that if a user will apply the metaheuristic for solving the same problem independently then what is the probability that she/he will find a result better than some threshold value. Alternatively, how many runs would be required to ensure that one of the runs has better performance than the median run reported by the developer of the metaheuristic. In this paper, an attempt has been made to provide a novel framework for testing metaheuristics that use some randomness in their search process. The paper advocates that the output of metaheuristics be treated as Random Variable and thus, apply confidence interval approach [11, 12] for providing performance guarantees.

The paper is further organized as follows: Sect. 2 discusses current methods of testing metaheuristics. Section 3 presents the proposed framework of testing metaheuristics. Testing and Analysis of the proposed framework of testing metaheuristics have been discussed in Sect. 4. The paper concludes with some open problems for future work in Sect. 5.

2 Related Work

Metaheuristics are often validated by solving a set of well-known benchmark problems [5]. The testing procedure involves executing the solver several times with an independent set of random numbers, which are often generated by using seed controlled pseudo-random number generators [13]. The descriptive statistics like Best, Mean, Median, Worst, Variance/Standard Deviation, Success Runs, etc. of objective function values along with number iterations/function evaluations for every problem is computed to showcase the working of Metaheuristic [9]. As metaheuristics promise good solutions so the new metaheuristic is compared against available metaheuristics to show that either it is better than existing ones or at least as good as the existing counterparts. The inferential statistic is used for performing comparative testing and so it is assumed that the outcome of metaheuristics, i.e., Objective function value or Number of Function Evaluations are Random Variables. The test statistics for Inferential testing depend on the knowledge regarding probability distribution of the outcome, and the number of samples. If the probability distribution is normal and population variance is known then parametric testing using z statistic can be performed. Alternately, if population variance is unknown then parametric testing using t-statistic can be performed [12]. However, if the probability distribution is unknown then either large samples can be collected and normality can be assumed using Central Limit Theorem [14] or we can use non-parametric tests [10]. However, power of non-parametric tests is far weaker as compared to parametric tests which lead to rejection of NULL Hypothesis at lesser confidence [12]. Further, no attempt has been made till date to provide performance guarantee for Metaheuristics either by using analytical techniques in deterministic sense [4] or by using confidence intervals in probabilistic sense. An attempt has been made in this paper to propose a framework that will help in providing performance guarantee through confidence intervals with some probability. Performance guarantees are important if metaheuristics are to be commercialized as products because when we buy products, we do not just compare the product against the competitors but also look for performance against the specifications. Further, quality testing of a product is performed against specifications and not against competing products.

3 Proposed Framework

The cornerstone of proposed framework of performance testing is to treat the final outcome of metaheuristics as a Random Variable, i.e., there is a true optimal value, which metaheuristic is trying to reach using a randomized process so, in every independent run, it produces an outcome that forms sample space which is closer to true optima value. We are assuming that metaheuristic under performance testing has been fine-tuned [15] and has produced near-optimal solutions in case of benchmarks problems with known optimum or solutions near the current known best solution in case of problems (Real-world as well as Benchmark) with unknown optimal solutions.

A random variable is characterized by its probability distribution function. If the probability distribution function of a random variable is known then the problem is trivial, but if the probability distribution function is to be estimated from the sample data then it becomes a challenge as the estimation of the probability distribution of a random variable with high accuracy is a difficult optimization problem [16]. However, in such cases Chebyshev's Inequality [12] can be used, which states that for any random variable X with mean μ and variance σ^2, for $k > 0$, the probability that sample average Xav is away from population mean μ by k times variance σ^2 is given below:

$$P\left(|Xav-\mu| > k\sigma^2\right) < 1/k^2 \qquad (1)$$

The proposed framework for performance testing of metaheuristics is as follows:

1. **INPUT: Tuned Metaheuristic(s) and Problem(s)**
2. **Run Experiments on each combination for N times (N>25)**
3. **Compute Descriptive Statistics**
4. **Estimate the distribution:**
    IF estimated distribution is accurate THEN
        GO TO Step 5
    ELSE IF Sample Size N is Large THEN
        Assume Normal Distribution and GO TO Step 5
    ELSE
        Use Chebyshev's Inequality and GO TO Step 5
5. **Compute Confidence interval at X% probability**
6. **Comparative study (Hypothesis Testing)**
    IF Distribution Known OR Sample Size is Large THEN
        Perform Parametric Hypothesis Testing
    ELSE
        Perform Non-Parametric Hypothesis Testing
7. **OUTPUT: Performance Guarantee on Output with X% probability AND Result of Comparative Testing**

In Step 1, Metaheuristic(s) and problem sets are selected and in Step 2 all meta-heuristic(s) solves all problem set(s) in N independent Runs to collect data for computing performance guarantee. In Step 3, descriptive statistics like Best, Worst, Median, Mean, and Standard Deviation are computed for collected data for each metaheuristic. In Step 4, the probability distribution for output of metaheuristic is determined, which, if known is an easy problem, but if unknown, then it becomes a difficult problem. If distribution is unknown then estimate the distribution by Maximum Likelihood Estimation method or similar methods [12]. Alternately if sample size is large then assume Normal distribution and estimate mean and variance from sample data and use t-statistics [12]. Another option is to use Chebyshev's Inequality if distribution cannot be estimated accurately. In step 5, compute the confidence interval using probability distribution function or Chebyshev's inequality at X% probability for one side/tail of the distribution as Metaheuristics solve either Maximization or Minimization problem so only side of the distribution is important because we can err on only one side depending on whether the problem being solved is Minimization or Maximization problem. In step 6, Comparative Testing between different Metaheuristics can be performed as given in literature [9, 10, 12] by using parametric or non-parametric tests. Parametric tests are more powerful but restrictive as compared to non-parametric tests. In step 7, we will get the performance guarantee on the Metaheuristic and will also know whether the metaheuristic is better than other competing metaheuristics.

4 Testing and Analysis

The proposed framework has augmented the existing method of testing and validating metaheuristics by introducing confidence interval [11], so the testing has been performed for proving our hypothesis that confidence intervals can be used for providing performance guarantee by taking metaheuristics and benchmark problems from the existing literature [9]. We have taken CEC-2017 real parameter benchmark optimization problems [9] and the metaheuristic chosen is from [17]. The data for five selected benchmark problems are shown in Table 1 with Confidence bound at 50, 75, 95, and 99% levels using Chebyshev's Inequality given in Eq. (1). Further, the same data is plotted in Fig. 1 to show the relative position of Best, Worst, Mean, Median, and Confidence bound at 50, 75, 95, and 99% levels. The Y-Axis represents Objective Function Values and X-Axis represents the problems.

It is evident from Table 1 and Fig. 1 that performance guarantee can be given with the help of Confidence Intervals as 99% bound is much above the worst performing run. In fact, 95% bound is also slightly above the worst performing run. The 50% bound is also above the Mean for all the five problems. We also performed Wilcoxon

Table 1 Descriptive statistics and confidence bounds

#	Best	Worst	Median	Mean	Std	50%	75%	95%	99%
f5	3.01E+01	6.00E+01	4.40E+01	4.39E+01	5.61E+00	5.18E+01	5.51E+01	6.90E+01	1.00E+02
f7	1.29E+02	1.60E+02	1.44E+02	1.45E+02	6.70E+00	1.54E+02	1.58E+02	1.75E+02	2.12E+02
f11	4.27E+01	3.04E+02	1.04E+02	1.13E+02	4.32E+01	1.74E+02	2.00E+02	3.06E+02	5.45E+02
f13	5.33E+01	2.51E+02	1.40E+02	1.45E+02	3.80E+01	1.99E+02	2.21E+02	3.15E+02	5.25E+02
f15	8.67E+01	2.62E+02	1.65E+02	1.62E+02	3.81E+01	2.16E+02	2.38E+02	3.32E+02	5.43E+02

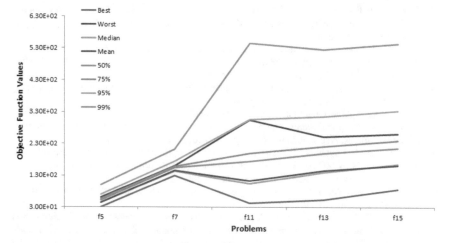

Fig. 1 Graph highlighting relative position of descriptive statistics measures and confidence bounds

Signed Rank Test to prove our claim on 29 problems [9, 17] with Null hypothesis being worst Objective Function Value μ_1 is greater than or equal to the 95% Confidence Interval μ_2, i.e., H_0: $\mu_1 \geq \mu_2$. An alternate hypothesis is H_1: $\mu_1 < \mu_2$. The result shows that null hypothesis can be safely rejected as Wilcoxon's Signed Rank test statistic is eight and less than the critical value of 140 at significance level of $\alpha = 5\%$, which indicates a similar result as that shown by descriptive statistics in Table 1. Further, it gives statistical stability that Confidence Interval method can provide performance guarantee for the above mentioned examples.

5 Conclusions and Future Work

Testing of Metaheuristics is an important research area, however, current methods of testing do not help in arriving at Performance Guarantee. An attempt has been made to provide Performance Guarantee through the application of Confidence Intervals using Chebyshev's Inequality in a probabilistic sense. Performance Guarantee will help in improving the acceptability of Metaheuristics in solving real-world optimization problems. However, there are open problems and areas of improvement like estimation of probability distribution from samples and computing power of test in non-parametric studies which will further help in providing tighter performance guarantee on Metaheuristics.

References

1. S. Shan, G.G. Wang, Struct. Multidisc. Optim. **41**, 219 (2010). https://doi.org/10.1007/s00158-009-0420-2
2. C. Blum, A. Roli, Hybrid metaheuristics: an introduction, in *Hybrid Metaheuristics. Studies in Computational Intelligence*, vol 114, ed. by C. Blum, M.J.B. Aguilera, A. Roli, M. Sampels (Springer, Berlin, Heidelberg, 2008)
3. A. Nakib, P. Siarry, Performance analysis of dynamic optimization algorithms, in *Metaheuristics for Dynamic Optimization. Studies in Computational Intelligence*, vol. 433, ed. by E. Alba, A. Nakib, P. Siarry (Springer, Berlin, Heidelberg, 2013)
4. M.-H. Lin, J.-F. Tsai, C.-S. Yu, A review of deterministic optimization methods in engineering and management. Math. Problems Eng. **2012** (2012)
5. B. Sarasola, E. Alba, Quantitative performance measures for dynamic optimization problems, in *Metaheuristics for Dynamic Optimization. Studies in Computational Intelligence,* vol. 433, ed. by E. Alba, A. Nakib, P. Siarry (Springer, Berlin, Heidelberg, 2013)
6. I. Boussaïd, J. Lepagnot, P. Siarry, A survey on optimization metaheuristics. Inf. Sci. **237**, 82–117 ISSN 0020-0255 (2013). https://doi.org/10.1016/j.ins.2013.02.041. http://www.sciencedirect.com/science/article/pii/S0020025513001588
7. K. Hussain, M.N. Mohd Salleh, S. Cheng et al., Artif. Intell. Rev. **52**, 2191 (2019). https://doi.org/10.1007/s10462-017-9605-z
8. M. Issa, A.E. Hassanien, I. Ziedan, Performance evaluation of sine-cosine optimization versus particle swarm optimization for global sequence alignment problem, in *Machine Learning Paradigms: Theory and Application. Studies in Computational Intelligence*, vol. 801, ed. by A. Hassanien (Springer, Cham, 2019)

9. N.H. Awad, M.Z. Ali, J.J. Liang, B.Y. Qu, P.N. Suganthan, Problem definitions and evaluation criteria for the CEC 2017 special session and competition on single objective bound constrained real-parameter numerical optimization. Technical Report, Nanyang Technological University, Singapore, Nov 2016

10. S. García, D. Molina, M. Lozano et al., J. Heuristics, A study on the use of non-parametric tests for analyzing the evolutionary algorithms' behaviour: a case study on the CEC 2005 special session on real parameter optimization. **15**:617 (2009). https://doi.org/10.1007/s10732-008-9080-4

11. S. Weerahandi, Generalized confidence intervals, in *Exact Statistical Methods for Data Analysis*. Springer Series in Statistics (Springer, New York, NY, 1995)

12. D.C. Montgomery, G.C. Runger, in *Applied Statistics and Probability for Engineers*, 5th edn. (Wiley, 2011), 765 pp

13. L. Rajashekharan, C. Shunmuga Velayutham, Is differential evolution sensitive to pseudo random number generator quality?—an investigation, in *Intelligent Systems Technologies and Applications*. Advances in Intelligent Systems and Computing, vol. 384, ed. by S. Berretti, S. Thampi, P. Srivastava (Springer, Cham, 2016)

14. H. Fischer, A history of the central limit theorem: from classical to modern probability theory, in *Sources and Studies in the History of Mathematics and Physical Sciences* (Springer, New York, 2011). https://doi.org/10.1007/978-0-387-87857-7

15. N. Mani, Gursaran, A.K. Sinha, A. Mani, Taguchi-based tuning of rotation angles and population size in quantum-inspired evolutionary algorithm for solving MMDP, in *Proceedings of the Second International Conference on Soft Computing for Problem Solving (SocProS 2012), December 28–30, 2012*. Advances in Intelligent Systems and Computing, vol. 236, ed. by B. Babu et al. (Springer, New Delhi, 2014)

16. R. Cacoullos, Estimation of a probability density. Ann. Inst. Stat. Math. (Tokyo) **18**(2), 179–189 (1966)

17. J. Brest, M.S. Maucec, B. Boskovic, Single objective real-parameter optimization: algorithm jSO, in *2017 I.E. Congress on Evolutionary Computation (CEC)* (2017), pp. 1311–1318

Automatic Clustering of Hyperspectral Images Using Qutrit Based Particle Swarm Optimization

Tulika Dutta, Sandip Dey and Siddhartha Bhattacharyya

Abstract Hyperspectral Images (HSI) contain a lot of data channels. Due to their high dimensionality, it is difficult to design systems which are able to find optimal number of clusters to segment them. A qutrit based particle swarm optimization (PSO) for automatic clustering of hyperspectral images is introduced in this paper. A Band Fusion Technique is implemented by improving the Improved Subspace Decomposition Algorithm using SF Index as the fitness function. A new method for maintaining the superposition state of the qutrits is also successfully designed. A comparison with classical PSO is performed to clearly establish the supremacy of the proposed technique with respect to Peak signal-to-noise ratio (PSNR), Jaccard Index, Sørensen-Dice Similarity Index and the computational time. Finally, the unpaired two-tailed t test is conducted between the proposed technique and classical PSO, which renders better results for proposed qutrit based technique. The experiments are carried out on the Salinas Dataset. The proposed technique yields better results in all the tests conducted in comparison to the classical PSO.

Keywords Automatic clustering · Hyperspectral image segmentation · Particle swarm optimization · Multilevel quantum systems · Qutrit

T. Dutta
University Institute of Technology, Burdwan University, West Bengal, India
e-mail: munai.tulika@gmail.com

S. Dey
Sukanta Mahavidyalaya, Dhupguri, Jalpaiguri, West Bengal, India
e-mail: dr.ssandip.dey@gmail.com

S. Bhattacharyya (✉)
RCC Institute of Information Technology, West Bengal, India
e-mail: dr.siddhartha.bhattacharyya@gmail.com

© Springer Nature Singapore Pte Ltd. 2020
S. Bhattacharyya et al. (eds.), *Intelligence Enabled Research*,
Advances in Intelligent Systems and Computing 1109,
https://doi.org/10.1007/978-981-15-2021-1_4

1 Introduction

Numerous researches are being carried out on hyperspectral images. HSI can be perceived as a stack of images. They are captured over a particular target area, with different wavelength intervals of spectral channels. HSI is cursed with high dimensionality and redundant data, causing it difficult to extract information from them. Hence, several fusion techniques are used to reduce the band numbers or choose minimal bands for applying them in image segmentation, classification [9], and other methods.

Image Segmentation is the process of isolating an image into non-contiguous parts based on some common traits like texture, color or pixel intensity. Clustering-based segmentation approach divides the image into various groups, referred to as clusters. It can be a tedious job to determine the number of clusters in some occasions. So automatic clustering of images has become an active research area [4]. K-means [5], Fuzzy C Means [1], etc. are few widely used techniques. Cluster Validity Indices (CVI) are used to measure the efficiency of clustering. Dunn's Index [8], I-Index [16] are few examples of CVI's.

Metaheuristics are iterative, stochastic algorithms, which can be effectively used in pursuance of finding the solutions to various optimization problems. Nature-inspired, mainly Biology-Inspired metaheuristic algorithms are proven to be more effective in finding near-optimal solutions. Some well known metaheuristic algorithms are Genetic Algorithm (GA) [3], Particle Swarm Optimization (PSO) [20], Ant Colony Optimization (ACO) [7], Differential Evolution (DE) [3], Simulated Annealing (SA) [3] to name a few.

Quantum Computing (QC) is a comparatively new computing paradigm based upon Richard Feynman's idea of building new generation computers. This is conceptualized on principles of quantum physics [17]. They produce much accelerated results than their classical counterparts. Researchers have lately introduced quantum-based metaheuristic algorithms for segmentation of images [2]. Quantum-inspired metaheuristics work on the basic principles of QC like *qubits* or *quantum bit* and *quantum superposition*. There exist few quantum-inspired algorithms in the literature, that are designed for bi-level/multilevel image thresholding of different images [3, 21].

The objective of this paper is to propose a qutrit based automatic clustering algorithm of hyperspectral images, by minimizing the number of bands. The band fusion technique lowers the space complexity of dealing with 150–220 bands present in each image. The modified improved subspace decomposition [22] technique is implemented, for choosing minimal number of bands by applying Shannon entropy [22]. *Qutrit* based population is employed, which reduces the space and time complexity to a greater extent. The SF Index [19] is employed for determining the optimal number of clusters [14]. SF Index [19] measures the efficiency of clustering. The data is assigned to the active cluster center with the highest fuzzy membership value.

The contributions of the proffered work are presented as follows:

- Band Fusion Technique of hyperspectral images to choose the minimal number of bands.
- A Qutrit based Automatic Cluster detection system is implemented for hyperspectral images.
- New Qutrit updation technique for maintaining Quantum Orthogonality at each iteration of the PSO.

2 Related Work

Numerous techniques are adapted for choosing the most discriminative bands. This is implemented by decreasing the correlation or by increasing the information between chosen bands. The most commonly used techniques in the literature are band ranking, band clustering, and subspace decomposition techniques [22].

PSO is a widely implemented optimization algorithm. This is inspired from the flocking behavior of birds or schooling behavior of fishes [20]. Till date, a lot of updated PSO's have been implemented since Dr. Kennedy and Dr. Eberhart first proposed it in 1995 [20]. In [15], a Dynamic PSO is fused with K-means algorithm for image segmentation. In [11], a Levy Flight PSO is implemented to improve the results from getting stuck into local optima.

Quantum-inspired evolutionary algorithms are extensively and successfully compared to their classical counterparts. In 1996, Narayanan and Moore introduced the concept of Quantum-Inspired Genetic Algorithm, which outperformed classical GA [18]. Various Quantum-Inspired Genetic Algorithms, PSO, DE, ACO, SA, etc, have been introduced in [2, 3]. A Quantum-Inspired Automatic Clustering Technique using ACO is proposed in [4].

Clustering techniques can be classified into two categories, viz., Hierarchical and Partitional Clustering. In [5], a new algorithm is proposed by fusing the K-means algorithm with a subtractive clustering algorithm. In [14], an Ant Colony based K-Harmonic-Means algorithm is presented.

Best segmentation can be evaluated using CVI. Several CVI are found in the literature. Mostly, all indices require at least two clusters to determine the compactness and separability of the clusters. Some widely used CVI's are Dunn's Index [8], I-Index [16], to name a few. The SF Index introduced in [19] is implemented in this work for cluster validation.

3 Some Basic Concepts

Some important concepts related to the proffered work are briefly discussed in the subsections presented below.

3.1 Quantum Computing

Quantum computing is derived from the field of Quantum mechanics, which is extensively applied to develop different algorithms [4, 21]. A *qubit* is the smallest unit of a quantum information system. It has two computational basis states, viz., $|0\rangle$ and $|1\rangle$ [17]. The qubits can exist in a state of a linear combination of the basis states, popularly known as quantum superposition. Formally, it can be represented as

$$|q\rangle = \alpha_0|0\rangle + \alpha_1|1\rangle \tag{1}$$

where, a strict normalization constraint is always followed. The normalization constraint can be stated as

$$\alpha_0^2 + \alpha_1^2 = 1 \tag{2}$$

The denouement of measurement of quantum state is either the *ground state* ($|0\rangle$) or the *excited state* ($|1\rangle$), which is basically a probabilistic measurement.

Instead of employing bi-level logic (having 2^2 in numbers) for determining a quantum states, if a multilevel logic is used for the same, it is called a *qudit*. The three-level quantum state or *qutrit* state is the most elementary multilevel quantum logic [21]. It contains the basis states, viz., $|0\rangle$, $|1\rangle$ and $|2\rangle$. The superposition state, in such case, can be expressed as

$$|q\rangle = \alpha_0|0\rangle + \alpha_1|1\rangle + \alpha_2|2\rangle \tag{3}$$

The normalization constraint is expressed as follows:

$$\alpha_0^2 + \alpha_1^2 + \alpha_2^2 = 1 \tag{4}$$

The matrix representation of *qutrit* states is given by

$$|0\rangle = \begin{pmatrix} 1 \\ 0 \\ 0 \end{pmatrix}, |1\rangle = \begin{pmatrix} 0 \\ 1 \\ 0 \end{pmatrix}, |2\rangle = \begin{pmatrix} 0 \\ 0 \\ 1 \end{pmatrix} \tag{5}$$

3.2 Hyperspectral Image Band Fusion Method

Due to the large number of contiguous spectral bands in hyperspectral images, redundancy may become a common problem. For image segmentation, dimensionality reduction is a necessary step in HSI processing. The improved subspace decomposition technique is employed in the method proposed by Xie et al. [22]. Using correlation coefficient (R_{ij}) and spectrogram of the image, b number of bands are divided into n number of non-overlapping groups. The correlation coefficient, R_{ij} between band b_i and b_j is written as

$$R_{ij} = \frac{\sum_{k1}^{n} \left(b_{ik} - \overline{b_i}\right) \left(b_{jk} - \overline{b_j}\right)}{\sqrt{\sum_{k=1}^{n} \left(b_{ik} - \overline{b_i}\right)^2 \sum_{k=1}^{n} \left(b_{jk} - \overline{b_j}\right)^2}} \tag{6}$$

Each cluster of bands has high R_{ij} value and it represents bands with similar features. The different groups have lower correlation and they signify presence of different features. Based on Shannon entropy, $E(b_i)$, one band is selected from each of the n groups. The information contained in the band (b_i) is quantified by the entropy definition as given by

$$E(b_i) = -\int_{b_i} p(b_i) \log p(b_i) db_i \tag{7}$$

where, $p(b_i)$ stands for probability density function of band b_i. The information content of each of the n bands, containing S number of elements, is calculated using the following polynomial

$$\frac{1}{S} \left(\sum_{j=1}^{S} E(b_j) \right) \tag{8}$$

Three bands with highest information are chosen from n groups, containing one band from each group, respectively.

4 Proposed Methodology

In this paper, an Automatic Clustering-based Qutrit version of Particle Swarm Optimization (AC-QuPSO) for HSI is proposed. First, the preprocessing of the hyperspectral images is performed to select the minimal number of bands [22]. The total number of bands are divided into groups using Eq. (6) and the spectrogram of the hyperspectral image cube. Using Eq. (8), the information contained in every band is calculated. Minimal bands having the highest information are chosen for the automatic clustering process.

A random population (P) of n particles is initially generated. The length of each string is chosen as m. As the initial base population doesn't contain any information, the qutrits are being initialized with equal values. The initialization is done in such a way that the quantum orthogonality is maintained using the following equation

$$|q\rangle = \frac{1}{\sqrt{3}}|0\rangle + \frac{1}{\sqrt{3}}|1\rangle + \frac{1}{\sqrt{3}}|2\rangle \tag{9}$$

Qutrit state observation, which transforms the quantum chromosome into its classical representation using numbers 0, 1, 2 for the 3 states is performed in this paper. The following procedure is used for obtaining the Qutrit State Observations [21].

– Pick a Qutrit p from (P),containing states α_0, α_1, α_2
  - rand $=$ random number in range $[0,1]$
  - If rand$<$ $[\alpha_0^2]$, then $c = 0$
  - Else If rand$<$ $[\alpha_0^2 + \alpha_1^2]$, then $c = 1$
  - Else $c = 2$

The following process is carried out for G_x generations. For each pixel (D), its distance to the active cluster centers is calculated using the membership criteria $m\left(\text{clust}_j/xd_i\right)$ [14], as given by

$$m\left(\text{clust}_j/xd_i\right) = \frac{\|xd_i - \text{clust}_j\|^{-p-2}}{\sum_{j=1}^{k}\|xd_i - \text{clust}_j\|^{-p-2}}, \quad m\left(\text{clust}_j/xd_i\right) \in [0, 1] \qquad (10)$$

Each point is assigned to the cluster center with the highest membership value.

Using SF index [19], the clusters' validity is calculated for determining the fitness of each cluster. The "between-class distance (bcd)" and "within class distance (wcd)" are used to determine the quality of clustering. Formally, SF index can be defined by

$$SF = 1 - \frac{1}{e^{e^{bcd-wcd}}} \qquad (11)$$

where,

$$\text{bcd} = \frac{\sum_{i=1}^{k}\|\text{clust}_i - \text{clust}_m\| \cdot n_i}{n \cdot k} \text{ and wcd} = \sum_{i=1}^{k}\left(\frac{1}{n_i}\sum_{xd_\in \text{clust}_i}\|xd_i - \text{clust}_i\|\right) \qquad (12)$$

Here, xd_i denotes the data points, clust_i denotes the ith active cluster center, clust_m is the centroid of all clusters, n_i denotes the number of data in ith cluster, n denotes the total data points and k denotes the number of clusters.

The following formula is used to update the velocity (v) of the particles:

$$v[t + 1] = W * v[t] + \rho1 * \text{rand}1 * (\text{pbest}[t] - \text{pop}[t]) + \rho2 * \text{rand}2 * (\text{gbest}[t] - \text{pop}[t]) \qquad (13)$$

The pop$[t]$ is taken as the corresponding amplitude of the quantum population. The global best is denoted by gbest, whereas the particle best is denoted using $pbest$.

The population is updated using the following formula

$$\text{pop}[t + 1] = \text{pop}[t] + v[t + 1] \qquad (14)$$

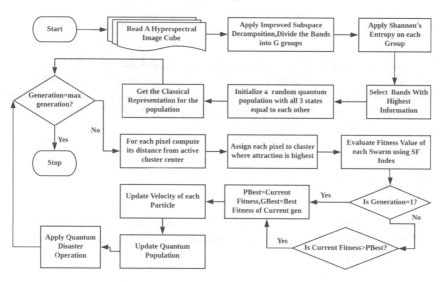

Fig. 1 Block diagram of AC-QuPSO

Assuming $pop[t] = \alpha_0$ as c $= 0$, then the remaining two amplitudes are updated using the concept of Quantum Orthogonality Principle Equation (4) as follows:

- As $c = 0$
- $\alpha_1 = \sqrt{\frac{(1-\alpha_0^2)}{2} - \text{random number}(< 0.1)}$
- $\alpha_2 = \sqrt{1 - (\alpha_0^2 + \alpha_1^2)}$

The Quantum Population easily gets stuck into local optima. To avoid this, a part of the population is re-initialised to the initial state. The members having least fitness values, are chosen. This operation is called the Quantum Disaster Operation. The optimum fitness values and the corresponding clustered image are obtained when the fitness function gets the highest value [19]. The proposed technique is diagramatically presented in Fig. 1.

5 Experimental Results

A brief description of the dataset is provided in Sect. 5.1. In Sect. 5.2, the details of the fitness function is discussed. In Sect. 5.3, the experimental results are noted in tabular forms along with resultant segmented images.

5.1 Dataset

The proposed AC-QuPSO is implemented on the Salinas Dataset [13]. This dataset was captured by AVIRIS sensor over Salinas Valley, California. The dataset contains 224 bands, with a resolution of $512 * 217$ pixels for each band. The spatial resolution is very high (3.7-m pixels). There are 16 classes in the ground truth image consisting of vegetables, bare soils, vineyard fields, etc.

5.2 Fitness Function

In the proposed technique, the SF cluster validity index [19], as given in Eq. (11), is used as the fitness function. The maximization of the SF value leads to the optimum number of clusters.

5.3 Analysis

The Ground Truth Image of Salinas Dataset [13], the fused image obtained using ISD [22] and the segmented images using AC-PSO and AC-QuPSO are presented, in Fig. 2. Three initial parameters, the inertia influence (W), weights for global best ($\rho1$) and local best ($\rho2$) are taken as 0.45, 0.55, 0.55 (for AC-PSO) and 0.45, 0.65, 0.85 (for AC-QuPSO). The best time along with the mean and standard deviations of the fitness values, PSNR [12], Jaccard Index [10] and Sørensen-Dice [6] similarity tests are recorded in Table 1. PSNR [12] indicates better segmentation for higher values (more than 1). Jaccard Index [10] and Sørensen-Dice [6] give a value between 0 and 1, where 1 indicates optimal segmentation. The values of PSNR [12], Jaccard Index [10] and Sørensen-Dice [6] are relatively low when compared to gray-scale images. The comparison of segmented images with ground truth images is responsible for producing such results. In all the three indices, AC-QuPSO yields better results compared to AC-PSO. Few favorable outcomes are recorded in Table 2, indicating the optimal number of clusters. The two-tailed t-test is carried out. A 5% confidence level is considered to evaluate the p-value in this test. This is recorded in Table 3. Here, AC-QuPSO provides better results than AC-PSO in all cases. The execution is conducted in MATLAB R2019a on an Intel(R) Core(TM) $i7$ 8700 Processor in Windows 10 environment.

Fig. 2 Ground truth, selected bands (ISD), AC-PSO segmentation ($k = 7$), AC-QuPSO segmentation ($k = 6$)

Table 1 Best time, mean, standard deviation (STD), PSNR, Jaccard Index (JI) and Sørensen-Dice Similarity Index (SDI) for AC-PSO and AC-QuPSO

Sr No.	Process	Best time	Mean	STD	PSNR	JI	SDI
1	AC-PSO	207.96	0.7924	0.0109	4.6346	0.4632	0.6332
2	AC-QuPSO	60.21	0.7946	0.0072	6.4895	0.5007	0.6673

Table 2 Five favorable outcomes of the dataset for AC-PSO and AC-QuPSO

	Salinas dataset			
Process	AC-PSO		AC-QuPSO	
Sr. No.	Cluster No.	Fitness value	Cluster No.	Fitness value
1	7	0.7890	6	0.7966
2	6	0.7828	6	0.7995
3	7	0.7871	6	0.7832
4	7	0.8109	6	0.7925
5	7	0.7921	6	0.8012

Table 3 Result of two-tailed (unpaired) t-test between AC-QuPSO and AC-PSO

Dataset	Two-tailed p value	Significance
Salinas	0.000176	Significant

6   Conclusion

A Qutrit based PSO is proposed for the automatic clustering of hyperspectral images in this paper. The prime aspect of this technique is that without previous knowledge, the optimal number of clusters in hyperspectral images is determined. The band fusion technique used in the proposed technique also minimizes and employs only the minimal bands that are required in the hyperspectral images. The SF Index is implemented to validate the optimal number of clusters. The effectiveness of the proposed technique is established by comparing it with its classical counterpart with regards to the fitness measures, computational time, mean, and standard deviation of the fitness. The supremacy of the proposed technique is verified using the two-tailed t-test. The superiority of AC-QuPSO has also been established with reference to PSNR, Jaccard Index and Sorensen-Dice similarity measures.

The qudit version of the automatic clustering technique of hyperspectral images with the help of different band fusion techniques can be a future research interest.

Acknowledgements This work was supported by the AICTE sponsored RPS project on Automatic Clustering of Satellite Imagery using Quantum- Inspired Metaheuristics vide F.No 8-42/RIFD/RPS/Policy-1/2017-18.

References

1. H. Cui, K. Zhang, Y. Fang, S. Sobolevsky, C. Ratti, B.K.P. Horn, A clustering validity index based on pairing frequency. IEEE Access **5**, 24884–24894 (2017)
2. S. Dey, S. Bhattacharyya, U. Maulik, New quantum inspired meta-heuristic techniques for multi-level colour image thresholding. Appl. Soft Comput. **46**, 677–702 (2016)
3. S. Dey, S. Bhattacharyya, U. Maulik, Efficient quantum inspired meta-heuristics for multi-level true colour image thresholding. Appl. Soft Comput. **56**, 472–513 (2017)
4. S. Dey, S. Bhattacharyya, U. Maullik, *Quantum-Inspired Automatic Clustering Technique Using Ant Colony Optimization Algorithm* (IGI Global, 2018), pp. 27–54 (chap. 2)
5. N. Dhanachandra, K. Manglem, Y.J. Chanu, Image segmentation using K-means clustering algorithm and subtractive clustering algorithm. Procedia Comput. Sci. **54**, 764–771 (2015)
6. L.R. Dice, Measures of the amount of ecologic association between species. Ecology **26**(3), 297–302 (1945)
7. M. Dorigo, M. Birattari, Ant colony optimization. IEEE Comput. Intell. Mag. **1**, 28–39 (2006)
8. J.C. Dunn, Well-separated clusters and optimal fuzzy partitions. J. Cybern. **4**(1), 95–104 (1974)
9. A. Elmaizi, H. Nhaila, E. Sarhrouni, A. Hammouch, C. Nacir, A novel information gain based approach for classification and dimensionality reduction of hyperspectral images. Procedia Comput. Sci. **148**, 126–134 (2019)
10. S. Fletcher, M.Z. Islam, Comparing sets of patterns with the Jaccard index. Australas. J. Inf. Syst. **22** (2018)
11. Y. Hariya, T. Kurihara, T. Shindo, K. Jin'no, L flight PSO, in *2015 IEEE Congress on Evolutionary Computation (CEC)* (2015), pp. 2678–2684
12. A. Hore, D. Ziou, Image quality metrics: PSNR vs. SSIM, in *2010 20th International Conference on Pattern Recognition* (2010), pp. 2366–2369. https://doi.org/10.1109/ICPR.2010.579

13. Hyperspectral Remote Sensing Scenes - Grupo de Inteligencia Computacional (GIC) (2019). http://www.ehu.eus/ccwintco/index.php?title=Hyperspectral_Remote_Sensing_Scenes [Online; Accessed 7 Oct 2019]
14. H. Jiang, S. Yi, J. Li, F. Yang, X. Hu, Ant clustering algorithm with k-harmonic means clustering. Expert Syst. Appl. **37**(12), 8679–8684 (2010)
15. H. Li, H. He, Y. Wen, Dynamic particle swarm optimization and k-means clustering algorithm for image segmentation. Optik **126**(24), 4817–4822 (2015)
16. U. Maulik, S. Bandyopadhyay, Performance evaluation of some clustering algorithms and validity indices. IEEE Trans. Pattern Anal. Mach. Intell. **24**(12), 1650–1654 (2002)
17. D. McMahon, *Quantum Computing Explained* (Wiley, Hoboken, New Jersey, 2008)
18. A. Narayanan, M. Moore, Quantum-inspired genetic algorithms, in *Proceedings of IEEE International Conference on Evolutionary Computation* (1996), pp. 61–66
19. S. Saitta, B. Raphael, I.F.C. Smith, A bounded index for cluster validity, in *Machine Learning and Data Mining in Pattern Recognition* (Springer, Berlin, 2007), pp. 174–187
20. Y. Shi, R. Eberhart, A modified particle swarm optimizer, in *1998 IEEE International Conference on Evolutionary Computation Proceedings. IEEE World Congress on Computational Intelligence* (Cat. No.98TH8360) (1998), pp. 69–73
21. T. Valerii, Quantum genetic algorithm based on qutrits and its application. Math. Probl. Eng. **2018**, 1–8 (2018). https://doi.org/10.1155/2018/8614073
22. F. Xie, F. Li, C. Lei, J. Yang, Y. Zhang, Unsupervised band selection based on artificial bee colony algorithm for hyperspectral image classification. Appl. Soft Comput. **75**, 428–440 (2019)

Multi-class Image Segmentation Using Theory of Weak String Energy and Fuzzy Set

Soumyadip Dhar and Malay K. Kundu

Abstract Segmentation of a multi-class image is a major challenging work in image processing. The challenge arises as the uncertainties occur in the segmentation process. Here we present a novel method based on the concept of weak string energy to manage the uncertainties in the segmentation process. The concept of the weak string is utilized to find the location of the boundaries accurately among the segments. The segments of an image are generated based on the energy function in the fuzzy set domain in the proposed method. The accurate segments are generated when the function attains its minimum value. The segments are generated from an image without any prior knowledge about the total count of segments. The performance of the method is verified experimentally using different datasets and it is found to be quite satisfactory compared to the state-of-the-art methods.

Keywords Image segmentation · Fuzzy set · Weak string · Energy function

1 Introduction

Segmentation of a multi-class image is a crucial task in the field of computer vision. The objective of image segmentation is to group the pixels depending on some features of the image. The features may be the gray value of a pixel, color, texture, etc. Within a segment, the variation among the pixels should be as low as possible. For accurate image segmentation, it is necessary to diminish the uncertainties along with the appropriate localization of the boundaries between the segments. The segmentation of an image has diverse applications in medical science, metal industry, agriculture, robotics, etc.

S. Dhar (✉)
RCC Institute of Information Technology, Kolkata, India
e-mail: rccsoumya@gmail.com

M. K. Kundu
Indian Statistical Institute, Kolkata, India
e-mail: malay@isical.ac.in

© Springer Nature Singapore Pte Ltd. 2020
S. Bhattacharyya et al. (eds.), *Intelligence Enabled Research*,
Advances in Intelligent Systems and Computing 1109,
https://doi.org/10.1007/978-981-15-2021-1_5

In the literature on can find various techniques for image segmentation. They are based on edge [1, 2], region [3, 4] thresholds [5–8] etc. Broadly the methods found in the literature can be grouped into two varieties [9]. They are parametric methods and non-parametric methods. In the parametric method, the distribution of the pixels in an image is assumed. Based on the assumption the classifications of the pixels are done. On the contrary, no such assumptions are made in non-parametric methods. Non-parametric methods are quite popular for image segmentation. None of the methods stated above can handle uncertainties in the segmentation process. So, the performances of the methods are still below the expectation.

For handling the uncertainties in the segmentation process, the fuzzy set proposed by Zadeh [10] is the most popular tool in the field of image processing [8, 11–14]. In the fuzzy domain, an entropy function or an objective function depending on the membership values (e.g., fuzzy c-means objective function) is minimized to generate the optimal segments or clusters. The functions represent the uncertainty in the segmentation process. The minimization of the functions minimizes the uncertainty in the segmentation process. The objective functions have some inherent limitations in it. The objective functions do not consider local information for boundary detection between the segments. They do not emphasize the boundary regions between the two segments. As a result localization of the boundaries between two segments is poor and the performances have sufficient room for improvement.

Given the above-mentioned problem, in this paper, we incorporate the theory of weak string for the proper segmentation boundary localization. The weak string theory was successfully employed by Zisserman [15] for visual reconstruction. The theory was utilized for discontinuity detection in visual reconstruction. We incorporate the theory in the fuzzy domain for boundary localization. Here, we propose an objective function in the fuzzy domain that incorporates the theory of a weak string. The minimization of the objective function diminishes the uncertainties in finding the segments.

Our contribution is that here we propose an objective function based on the energy of weak string in the fuzzy domain. The function represents the ambiguities (gray level and spatial) in an image by incorporating the boundary information. Moreover, the segments are generated from an image without any prior knowledge about the total count of segments by a greedy iterative algorithm.

The paper is organized as follows: The theory of weak continuity constraints is explained in Sect. 2. The proposed segmentation using weak continuity constraints and the algorithm is described in Sect. 3. Section 4 describes the result and performance comparison.

2 Theory of Weak String

The theory of the weak elastic string means under the influence of weak continuity constraints [15] the nature of the weak string. The string is "weak" since it is breakable at some points with the violation of certain constraints. The theory tells that a string tries to remain in a low energy state. A string at a stretching position will remain continuous until the weak continuity constraints are violated. If the energy of the string crosses maximum permissible limit with the violation of constraints the string breaks to retain the lower energy. But, the breaking comes with the penalty at the point of discontinuities. The behavior of the weak string is represented by the following energy function E_{st}. The energy function is given by

$$E_{st} = D_{st} + S_{st} + P_{st} \tag{1}$$

where $D_{st} \sum_{i=1}^{N} (u_{st_i} - d_{st_i})^2$ represents the measure of faithfulness, $S_{st} = \lambda_s^2 \sum_{i=1}^{N-1}$ $(u_{st_i} - u_{st_{i+1}})^2 (1 - l_{st_i})$ represents the stretching energy and $P_{st} = \gamma_s \sum_{i=1}^{N-1} l_{st_i}$ is the penalty term due to break in the string. u_{st_i} is the stretched value of the string d_{st_i} at position i. l_{st_i} is the boolean value. λ_{st} is the scale and γ_{st} is the penalty parameter. At the point of discontinuity $l_{st_i} = 1$, and in the other place $l_{st_i} = 0$. In visual reconstruction, the objective is to find the minimum energy E_{st}, to fit the data d_{st_i} to u_{st_i}.

3 Multi-class Image Segmentation and the Concept of Weak String

3.1 Process of Image Segmentation

In the method of image segmentation, the objective is to cluster a set of pixels such that they are represented by a prototype. The aim is to reduce the total sum of the distance of the pixels from the prototypes in which the pixels belong. In clustering-based image segmentation, this is the main intuition. That means in the clustering the pixels in a cluster R_i where $i = 1, 2, \ldots, n$ are represented by a single cluster prototype v_i. The aim is to diminish the total sum of the distance of the pixels from the cluster prototype. The cluster should be such that the variation among the pixels in a cluster should be as low as possible. Moreover, the variation among the pixels of the different clusters should be as high as possible. The gradient would be high at the boundaries between two clusters. Thus, the localization of the boundaries is necessary for proper segmentation.

3.2 Segmentation Based on Fuzzy Set

In the fuzzy set-theoretic approach, the pixels are mapped into the fuzzy domain for clustering. In the proposed method we map the pixels into the fuzzy domain by a threshold t. The membership function for the pixels is given by the restricted equivalence function (REF) [12]. They are given by for an image $I(x), x \in [0 \quad L]$

$$\mu(x) = \begin{cases} \left(REF\left(\frac{x}{L-1}, \frac{m_b(t)}{L-1}\right)\right) & \text{if } x \leq t \\ \left(REF\left(\frac{x}{L-1}, \frac{m_o(t)}{L-1}\right)\right) & \text{if } x > t \end{cases} \tag{2}$$

Here, $REF(x, y) = 1 - |x - y|^k$ and $k > 0$. The REF was successfully used in the literature for image thresholding. The region where $x < t$ is the background and the region where $x >= t$ is the object. Here, m_b and m_o represent the cluster centers of background and object, respectively.

The threshold is used here for object background segmentation. That means we have to find out the proper t for which proper segmentation is generated. It is to be noted that the boundary between the two segments is given by neighboring points x, y where $(x <= t)\&(y > t)$ or $(x >= t)\&(y < t)$. Otherwise, x will not be the boundary point. So, the proper localization of the segmentation boundary requires neighborhood information. In the neighborhood information when the gradient is relatively high, the segmentation boundary is localized. So, the localization of the boundary means discontinuity detection in the pixels values. To incorporate the concept of the boundaries with the neighboring information, the concept of the weak string is utilized.

3.3 Proposed Concept of Weak String for Multi-class Segmentation

According to the theory of the weak string, a string always maintains a stable or low energy position. At the stretching position either is it continuous or it may break with the violation of weak continuity constraints. At the breaking position, the discontinuity is localized. But, the discontinuity comes with a penalty. In the stretching position if the energy is lower with the breaking than that of the continuous position, then the string will be broken. Otherwise, the string will remain continuous. We use the concept for segmentation in our method. For the segmentation process, we propose an energy function based on the theory of the weak string. The energy is given by

$$E = D + S + P \tag{3}$$

where

$$D = \sum_{i=1}^{c} \sum_{x \in R_i} (x - v(R_i))^2$$

$$S = \lambda^2 \sum_{i=1}^{c} \sum_{x \in R_i} \sum_{y \in N_x} (\mu(x) - \mu(y))^2 (1 - l(x))$$

$$P = \alpha \sum_{i=1}^{c} \sum_{x \in R_i} l(x) \tag{4}$$

In the above equation λ is the scale in which we take the neighborhood information. The large value of λ may mix the different texture of segments and small value can not capture the textures properly. In the proposed method the scale is taken as 7×7 overlapping window. The penalty α is high when the gradient is low and vice versa. Here we have taken $\alpha = \frac{1}{g(x)+1}$, where $g(x)$ becomes the gradient at x. When the gradient is low penalty is high and vice versa. $v(R_i)$ is the cluster prototype of cluster R_i. N_x is the set of neighborhood points of x. Proper boundary localization minimizes the uncertainties in segmentation. Thus, the minimization of E minimizes the uncertainties in segmentation and generate proper segments.

3.4 Proposed Algorithm

In the proposed algorithm the segmentation is executed automatically and it has no previous knowledge about the count of the segments in an image. For generating the proper threshold t the energy E is minimized. A threshold t divides an image into background and object. Then the portion in which E is maximum is taken for further segmentation. In this way, iteration will go on until there is no change in the threshold value. The steps of the algorithm are shown below

Step 1 **Image mapping into a fuzzy set**—In this step, the input image is trans-formed into a fuzzy set with the help of threshold t.

Step 2 **Minimization of the energy function**—In this step threshold t is found out for which energy $E = E_b + E_o$ is minimum. Here, E_b is the background energy and E_o is the energy of the object.

Step 3 **Compare the background and object energy**—The background energy E_b is compared with object energy E_o. If $E_o > E_b$ repeat the step 1 to step 2 for the object. Otherwise, repeat it for the background.

Step 4 **Continue the iterations**—The step 1 to step 3 are continued until no new thresholds are generated.

Step 5 **Generation of segments**—In this step multi-class image segments are generated depending on the thresholds.

4 Results and Discussion

We applied the proposed method on different types of images and verified the perfor-mances. The images included synthetic image, non-destructive testing (NDT) images [16] and natural images [17]. NDT images included eddy current, metal images, etc. In the industry non-destructive testing is important testing. Proper segmentation is necessary for finding an NDT image fault. The size of the various images used for the experiments varied from 200×200 to 300×300.

The proficiency of the proposed segmentation method was compared with state-of-the-art methods like Tizhoosh [18], Dhar [19], Naidu [8] and Pham [14]. The performance of the proposed method and all other methods were measured quantitatively by Segmentation Accuracy (S.A) [20] and Fuzzy evaluation (FE) [21]. The SA is the division of accurately classified pixels by all the pixels. It is written as

$$SA = \sum_{i=1}^{c} \frac{Al_i \cap G_i}{\sum_{j=1}^{c} G_j} \tag{5}$$

In the above equation, c becomes the total number of clusters. Al_i is the set of pixels belonging to the ith class generated by the proposed algorithm and G_i is the number of the pixels in the ground truth. In FE, a fuzzy-based approach was used for performance measure.

The qualitative performances of the method proposed here are shown in Fig. 1. From the figure, it is clear that the method could accurately select the thresholds for segmentation. The minimization of the energy function generated the thresholds properly. As already mentioned that the accurate localization of the segmentation boundary is necessary for the proper generation of the segments. The generation of the segmentation boundary helped to minimize the uncertainties in the segmentation process. The string theory in the fuzzy domain generated the segments properly. The segmentation of the images was executed iteratively depending on the energy of each segment. Thus, it is a greedy method for finding the proper number of the segments in an image. From the results on NDT images (row 2 and columns (c) and (g)) it is clear that the proposed method could identify the regions under a highly uncertain environment. The NDT images possessed a highly uncertain image pattern due to its unimodal or multimodal histograms. Most of the camera captured natural images also possessed highly uncertain image patterns due to their unimodal or multimodal histograms. From the qualitative results, we can say that the proposed method could handle the uncertainties in the images efficiently.

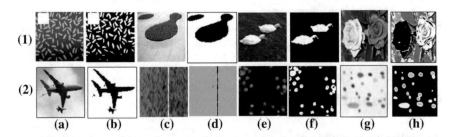

Fig. 1 Performance of the proposed method on different images from different datasets. Column-wise: **a**, **c**, **e** and **g** are original test images. **b**, **d**, **f** and **h** are corresponding segmentation results

Table 1 Quantitative performance on the natural images from [17]. ↑ signifies greater values becomes better achievement indicator

Image	Tizoosha	Dhar	Naidu	Pham	Proposed
SA↑	0.8342	0.8425	0.8429	0.8511	0.8632
FE ↑	0.8224	0.8402	0.8431	0.8443	0.8991

The proposed method was quantitatively compared with the other methods mentioned here. The results can be visualized from Table 1. Here, it is clear that the proposed method gave better performance than that of the other methods. Tizhoosh and Dhar used a fuzzy type-2 set-based method for image segmentation. But, the methods had no provision for boundary information. The weak string theory-based boundary detection in the proposed method helped to detect the segments accurately. The energy was minimized only when the proper boundary was detected. The method in Naidu [8] used the entropy-based segmentation technique. But, it could not use the local information properly. The method in [14] used the spatial information for the clustering of images. But, unlike the proposed method the method had no separate provision for boundary information. Moreover, for the methods, the number of classes were known in advance. In the proposed method the clusters were detected automatically without any external intervention.

5 Conclusion

Uncertainty handling in image segmentation is a big challenge. The existing measures of uncertainties in the fuzzy domain do not take into account the boundary between the segments. In the proposed method we overcome the limitation by introducing a weak string energy-based objective function in the fuzzy domain. The objective function takes into account the boundary information by a penalty parameter. The spatial information for segmentation is also taken into account by the scale parameter and neighborhood information. The uncertainties in the segmentation process are minimized when the energy function is reduced. The method is tested on various images and the performance is found to be better when compared to state-of-the-art methods. Currently, we are trying to extend the method for complex textured images.

References

1. A. Pratondo, C.-K. Chui, S.-H. Ong, Robust edge-stop functions for edge-based active contour models in medical image segmentation. IEEE Signal Process. Lett. **23**(2), 222–226 (2016)
2. C. Liu, W. Liu, W. Xing, An improved edge-based level set method combining local regional fitting information for noisy image segmentation. Signal Process. **130**, 12–21 (2017)

3. S. Niu, Q. Chen, L. De Sisternes, Z. Ji, Z. Zhou, D.L. Rubin, Robust noise region-based active contour model via local similarity factor for image segmentation. Pattern Recognit. **61**, 104–119 (2017)
4. Y. Haiping, F. He, Y. Pan, A novel region-based active contour model via local patch similarity measure for image segmentation. Multimed. Tools Appl. **77**(18), 24097–24119 (2018)
5. S. Borjigin, P.K. Sahoo, Color image segmentation based on multi-level tsallis-havrda-charvát entropy and 2D histogram using pso algorithms. Pattern Recognit. (2019)
6. J. Chen, B. Guan, H. Wang, X. Zhang, Y. Tang, W. Hu, Image thresholding segmentation based on two dimensional histogram using gray level and local entropy information. IEEE Access **6**, 5269–5275 (2018)
7. X. Zheng, H. Ye, Y. Tang, Image bi-level thresholding based on gray level-local variance histogram. Entropy **19**(5), 191 (2017)
8. M.S.R. Naidu, P.R. Kumar, K. Chiranjeevi, Shannon and fuzzy entropy based evolutionary image thresholding for image segmentation. Alex. Eng. J. **57**(3), 1643–1655 (2018)
9. S. Dhar, M.K. Kundu, Interval type-2 fuzzy set and theory of weak continuity constraints for accurate multi-class image segmentation. IEEE Trans. Fuzzy Syst. (2019). https://doi.org/10.1109/TFUZZ.2019.2930932
10. L.A. Zadeh, Quantitative fuzzy semantics. Inf. Sci. **13**, 159–176 (1971)
11. O.J. Tobias, R. Sear, Image segmentation by histogram thresholding using fuzzy sets. IEEE Trans. Image Process. **11**(12), 1465–1467 (2002)
12. H. Bustince, E. Barrenechea, M. Pagola, Image thresholding using restricted equivalence function and minimizing the measures of similarity. Fuzzy Sets Syst. **158**, 496–516 (2007)
13. M. Gong, Y. Liang, J. Shi, W. Ma, J. Ma, Fuzzy c-means clustering with local information and kernel metric for image segmentation. IEEE Trans. Image Process. **22**(2), 573–584 (2013)
14. T.X. Pham, P. Siarry, H. Oulhadj, Integrating fuzzy entropy clustering with an improved PSO for MRI brain image segmentation. Appl. Soft Comput. **65**, 230–242 (2018)
15. A. Blake, A. Zisserman, Weak continuity constraints in computer vision. Intern. Rep. (1986)
16. M. Sezgin, B. Sankur, Survey over image thresholding techniques and quantitative performance evaluation. J. Electron. Imaging **13**(1), 146–165 (2004)
17. http://www.wisdom.weizmann.ac.il
18. H. Tizhoosh, Image thresholding using type-2 fuzzy sets. Pattern Recognit. **38**, 2363–2372 (2005)
19. S. Dhar, M.K. Kundu, A novel method for image thresholding using interval type-2 fuzzy set and bat algorithm. Appl. Soft Comput. **63**, 154–166 (2018)
20. C. Li, R. Huang, Z. Ding, J.C. Gatenby, D.N. Metaxas, J.C. Gore, A level set method for image segmentation in the presence of intensity inhomogeneitics with application to MRI. IEEE Trans. Image Process. **20**(7), 2007–2016 (2011)
21. B. Ziółko, D. Emms, M. Ziółko, Fuzzy evaluations of image segmentations. IEEE Trans. Fuzzy Syst. **26**(4), 1789–1799 (2018)

Stabilization of Nonlinear Optimum Control Using ASRE Method

Nirmalya Chandra and Achintya Das

Abstract A Method to stabilize the nonlinear system using the Approximating Sequence of Riccati Equation (ASRE) is explained in this paper. For moderating the nonlinear disturbance a time-varying performance index is executed. The Dynamic State Transition Matrix is operated based on boundary condition and the response is optimized with affine input in the system. The Synchronization deviation between desired tracking response and actual response is manipulated to represent the linear form of output to the nonlinear system. The ASRE method describes the upgradation form of performance measuring parameter in nonlinear system with modified optimal input. In this formation, the characteristic of the positive definite controller matrix is controlled by differentiating it with respect to time. The stabilization of the output response and update optimum input is represented by a quadratic regulator based on this controller matrix.

Keywords Non quadratic performance index · Synchronization deviation · Positive definite matrix · ASRE methodology

1 Introduction

The foundation of continuous time system tries to exhibit stability in response. It is very cumbersome task to find out the uncontrolled parameters in time-varying nonlinear system. Linear Quadratic Regulator (LQR) is generally used for Linear Dynamical System [1]. In Nonlinear system the LQR is applied by pretending the system parameters are linear [2]. The development of nonlinear Optimum Control around the year 1950s and 1960s has been addressed in various theoretical and

N. Chandra (✉)
Maulana Abul Kalam Azad University of Technology, Haringhata 741249, India
e-mail: chandra.nirmalya@gmail.com

A. Das
Kalyani Government Engineering College, Kalyani, Nadia 741235, India
e-mail: achintya.das123@gmail.com

© Springer Nature Singapore Pte Ltd. 2020
S. Bhattacharyya et al. (eds.), *Intelligence Enabled Research*,
Advances in Intelligent Systems and Computing 1109,
https://doi.org/10.1007/978-981-15-2021-1_6

practical aspects where the objective is to minimize the cost given by performance index [3].

The Dynamic Programming Principle (DPP) by Bellman in USA was going ahead towards some kind of Partial Differential Equation (PDE) [4]. The solution of this Differential Equation was done by Renowned Scientists Hamilton, Jacobi, and Bellman (HJB) and the process was notified as HJB equations [5, 6].

The Performance analyzer, i.e., performance index in time-varying system has moving initial and terminal time but the time interval between these two times is finite [7]. Here the oscillations of responses of this system, converge to nonlinear system is at a marginal level. So though the system is not in stable condition, the characteristic of system response may not be hampered. The LTV system constructs an approximation solution named as Approximating Sequence of Riccati Equation (ASRE) [8, 9, 10].

The nonlinear system with linear input (affine system) is controlled with this method. In Sect. 2, The Linearization of Nonlinear System is analyzed.

In Sect. 3, the development of Non-Quadratic Performance Index is discussed which shows a form to represent the cost function in nonlinear system. The stabilization of the system by ASRE method is derived in this Section. Section 5, The Quadratic Control of a Time-Varying System is applied.

2 Linearization of Output to the Non-linear System

The deviation in between actual response ($y(t)$) and desired tracking response ($r(t)$) is known as Synchronization deviation. This is obtained as:

$$\varepsilon = |r(t) \sim y(t)| \tag{1}$$

For closed-loop the dynamical error,

$$\varepsilon = |y(t)| \tag{2}$$

Now, consider the synchronization output with controller gain 'K' is expressed as:

$$\left| \overset{\bullet}{y(t)} \right| = F(x(t) + G(x(t), u(t)) \tag{3}$$

is a nonlinear system with affine input.

Here, $F \in \mathbf{R}^{n \times n}$ and $G \in \mathbf{R}^{n \times m}$ are two positive definite matrices.

So from Eq. (1) and Eq. (2)

$$
\begin{aligned}
\dot{\varepsilon} &= F(y(t)) + G(y(t), u(t)) - F(r(t)) \\
&= F(\varepsilon(t)) + G((x_1(t) + x_2(t)u(t)) - F(r(t)) \\
&= F(\varepsilon(t)) + GK\varepsilon(t) - F(r(t)) \\
&= F_K(t)\varepsilon(t) - F(r(t))[F_K = (F + GK)\varepsilon(t)] \\
&\quad (\text{put}, ((x_1(t) + x_2(t)u(t)) = K\varepsilon(t)) \quad (4)
\end{aligned}
$$

In Eq. (4) the representation of error is in a linear form where actual output follows the input by putting $r(t) = 0$.

3 Example: A Nonlinear Self Explanatory System

Consider the optimal tracking state response of nonlinear system as:

$$
\frac{dx_1(t)}{dt} = x_2
$$
$$
\frac{dx_2(t)}{dt} = -a \cdot \cos(x_1) + \tau, \quad (5)
$$

where 'τ' is constant.

Now the optimal desired response set the value as: $\begin{bmatrix} r_1(t) \\ r_2(t) \end{bmatrix} = \begin{bmatrix} \pi/4 \\ 0 \end{bmatrix}$

To explore this problem, we set,

$$
\dot{y} = F(x_1(t)) + G(x_2(t))u(t) \quad (6)
$$

where $F = \begin{bmatrix} 0 & 1 \\ 0 & 0 \end{bmatrix}$ and $G = \begin{bmatrix} 0 \\ 1 \end{bmatrix}$

and $y_1(t) - a * \cos(x_1)$ and $y_2(t) = 1$. By pole placement technology at -1 and -2, the controller K is set as $K = [-2 \ -3]$, then the optimal input is written as

$$
u = a \cdot \cos(x_1) - 2(x_1 - \pi/4) - 3x_2 \quad (7)
$$

Using Matlab, the characteristics of system state and controller input are shown in Figs. 1 and 2.

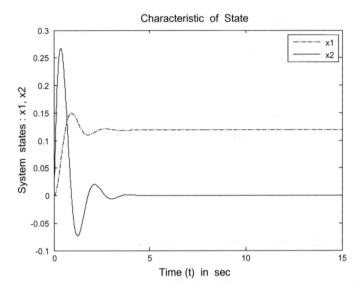

Fig. 1 Characteristic of state of the system

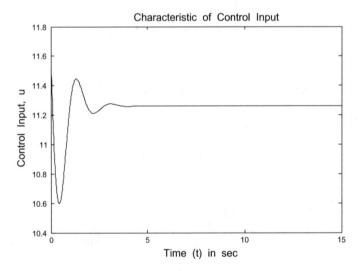

Fig. 2 The characteristic of control input

4   Stabilization of the System Using ASRE Methodology

Approximating Sequence of Riccati Equation (ASRE) control strategy keeps the deviation between desired output and actual output at sustainable level by Non-Quadratic and Time-Varying Performance Regulator.

Consider, $x(t)$, $u(t)$ represent the state vector and the plan input, respectively, and y be the system output. Now, the state and the plant input are regulated by nonlinear valued function '$M \in \mathbf{R}^{n \times n}$' and '$N \in \mathbf{R}^{n \times m}$'.

For the Non-Quadratic problem, the LTV is introduced as

$$\frac{dx^{\Delta}(t)}{dt} = M(x^{\Delta}(t) - \Delta x(t))x^{\Delta}(t) + N(x^{\Delta}(t), u^{\Delta}(t))$$

[where, $\left(x^{\Delta}(t) = x(t) + \Delta x(t)\right)$ and $\left(u^{\Delta}(t) = u(t) + \Delta u(t)\right)$]

$$y^{\Delta}(t) = \gamma\left[x^{\Delta}(t)\right] = \gamma(x(t) + \Delta x(t)) \tag{8}$$

where the elementary value of the state $x^{\Delta}(t_0) = x_0$,

The performance measuring cost function is given by

$$J^{\Delta}(u) = U_1\left|\varepsilon^{\Delta}(t_f)\right|^2 + U_2 \int_{t_0}^{t_f} [\varepsilon^{\Delta^T}(t)C\varepsilon^{\Delta}(t) + u^{\Delta^T}(t)Du^{\Delta}(t)]dt \tag{9}$$

where $u^{\Delta}(t) = -D^{-1}N^T(x^{\Delta}(t) - \Delta x(t))[P^{\Delta}(t)x^{\Delta}(t) - s^{\Delta}(t)]$ and $\varepsilon^{\Delta}(t) = r(t) - \gamma(x^{\Delta}(t))x(t)$

Here, in Eq. (9) $C \in \mathbf{R}^{n \times n}$ positive semi-definite matrix, $D \in \mathbf{R}^{m \times m}$ positive definite matrix, and $P^{\Delta} \in \mathbf{R}^{n \times n}$ positive definite matrix,

In the above Eq. (9) $s \in \mathbf{R}^n$ is transformation vector forward direction.

The vector $s^{\Delta}(t_f) \in \mathbf{R}^n$ is the solution of terminated point of nonlinear differential equation and it is expressed as

$$s^{\Delta}(t) = \gamma(x^{\Delta}(t_f) - \Delta x(t_f))Ur(t_f) \tag{10}$$

The positive definite matrix, $P^{\Delta}(t_f) = \gamma^T(x^{\Delta}(t_f) - \Delta x(t_f))U\gamma(x^{\Delta}(t_f) - \Delta x(t_f))$

Now, The change of state in an optimal tracking system stabilizes the state which is expressed by Linear Differential Equation as

$$\frac{dx^{\Delta}(t)}{dt} = N(x^{\Delta}(t) - \Delta x(t))D^{-1}N^T(x^{\Delta}(t) - \Delta x(t))s^{\Delta}(t) + M(x^{\Delta}(t)$$
$$- \Delta x(t)) - N(x^{\Delta}(t) - \Delta x(t))$$
$$\times [D^{-1}N^T(x^{\Delta}(t) - \Delta x(t))P^{\Delta}(t)] \tag{11}$$

5   Quadratic Control of a Time-Varying System

Consider a state variable time-varying system and output of this system are written as

$$\frac{dx_1(t)}{dt} = x_2(t)$$
$$\frac{dx_2(t)}{dt} = -x_1(t) + x_2(t).(1.5 - 0.2x_2^2(t)) + u(t) \quad \text{and} \quad y(t) = f(Cx(t) + Du(t))$$

$$(12)$$

The initial value of the state is $x_1(0) = x_2(0) = -5$ and C is Identity matrix and D set as zero. The system is represented in state matrix form as

$$\frac{dx(t)}{dt} = \begin{bmatrix} 0 & 1 \\ 1 & 1.5 - 0.2x_2(t) \end{bmatrix} \quad (13)$$

From the below Fig. 3, it is shown that the given states are oscillated with the propagation of time.

According to Matrix Differential Equation of the Algebraic Riccati Equation is

$$\frac{dP^{\Delta}}{dt} = P^{\Delta}(t)M(\cdot) - M^T(\cdot)P^{\Delta}(t) + P^{\Delta}(t)\gamma D^{-1}\gamma^T P^{\Delta}(t) - Q \quad (14)$$

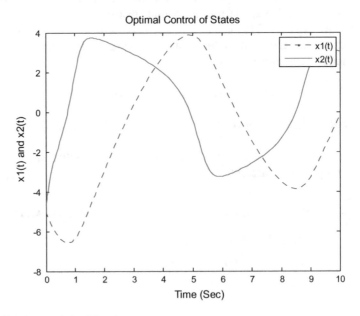

Fig. 3 The characteristic of the given two states

Thus Eq. (14) is called Differential Matrix Riccati Equation. The Characteristic of "$P^\Delta(t)$" is shown in the below Fig. 4 on the basis of the value of the Controller parameter (K) and "P". (Here, $P^\Delta(t) =$ in MATLAB syntax—P).

The value of the parameters—K and P is evaluated from MATLAB as (Table 1 and Fig. 4).

Using Quadratic Regulator the state and the output response are controlled. The following figures (Fig. 5a, b) show the regulable response of state and output and corresponding Optimum Input. The Characteristic of Optimum Input from Fig. 5b shows the optimum value of the input at initial point and the output and state response are saturated within the optimal point which is nearest to zero value. The state response starts with its initial value $|-5|$. The output also follows the characteristic of state. So after getting the optimum point from the Fig. 4 by ASRE solution, the saturation of States and Output is controlled through the system is oscillated at the early stages (Fig. 3).

Table 1 Value of controller parameter and definite matrix

K =	P =
0.7275	0.4077 −0.0293
0.2752	−0.0293 0.1112

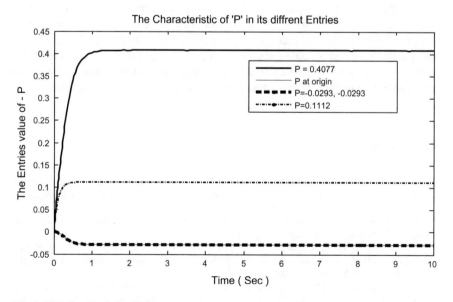

Fig. 4 The characteristic of "P"

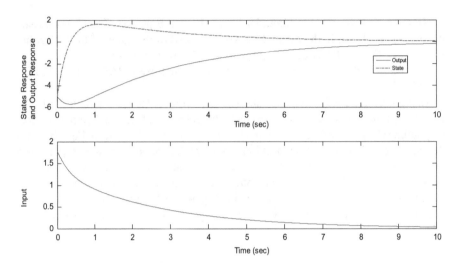

Fig. 5 **a** The controlled response of states and output of the system; **b** the characteristic of optimum input

6 Conclusion

In this paper, a new method to evaluate a performance analysis of nonlinear system using quadratic regulator is explained. By which the closed-loop dynamical error is minimized with desired optimal tracking response. The evaluation of ASRE control design is implemented in nonlinear affine system. The superiority of ASRE controller provides significant performance in real time application of nonlinear system. The extension of this method in non affine system is yet to be developed in future.

References

1. S.P. Banks, On the optimal control of nonlinear systems. Syst. Control Lett. **6**, 337–343 (1986)
2. M. Athans, P.L. Falb, *Optimal Control: An Introduction to the Theory and Its Applications* (McGraw-Hill, New York, 1966)
3. T. Cimen, Survey of state—dependent Riccati Equation in non linear optimal feedback control synthesis. J. Guid. Control Dyn. **35**(4), 1025–1047 (2012)
4. A.K. Nanda Kumaran, Calculus of variation to optimal control. Lecture notes at *Winter School on Stochastic Analysis and Control of Fluid Flow* (Department of Mathematics, IISER, Trivandrum, 3–20 December, 2012), pp. 1–23
5. B.D.O. Aderson, J.B. Moore, *Optimal Control* (Prentice Hall of India Private Limited, 2004)
6. H.J. Sussmann, J.C. Willems, 300 years of optimal control: the Brachystochrone to the maximum principle, in *History Session of the 35th Conference on Decision and Control* (Kobe, Japan, 1996), pp. 32–44
7. O. Hugues-Salas, S.P. Banks, Optimal control of chaos in nonlinear driven oscillators via linear time-varying approximations. Int. J. Bifurcation Chaos **18**(11), 3355–3374 (2008)

8. S.P. Banks, Exact boundary controllability and optimal control for a generalized Korteweg de Vries equation. Int. J. Nonlinear Anal., Methods Appl. **47**, 5537–5546 (2001)
9. S.P. Banks, Nonlinear delay systems, Lie algebras and Lyapunov transformations. IMA J. Math. Control. Inf. **19**, 59–72 (2002)
10. J. Balaram, G.N. Saridis, Design of sub optimal regulators for nonlinear systems, in *Proceedings of 24th IEEE Conference on Decision and Control* (Ft. Lauderdale, Fl, 1985), pp. 273–278

Patient-Specific Seizure Prediction Using the Convolutional Neural Networks

Ranjan Jana, Siddhartha Bhattacharyya and Swagatam Das

Abstract An epileptic seizure is a disease of the central nervous system caused by abnormal activity generated by neurons in the brain. Seizure reduces the quality of life of epileptic patients due to unconsciousness. In this paper, an efficient seizure prediction system is proposed to improve the quality of life. The raw EEG signal is converted into the EEG signal image. Then, a convolutional neural network is used for training the prediction system. The performance of the proposed system is evaluated using the CHB-MIT dataset. The classification accuracy of interictal and preictal states is achieved up to 94.33% using 10-fold cross-validation. Due to the presence of noise in the EEG signal, a pool based technique is used to make the decision on the majority of a 1 min EEG signal that increase the accuracy of the prediction of upcoming seizures.

Keywords CHB-MIT dataset · Convolutional neural network · EEG signals · Epileptic patient · Seizure prediction

1 Introduction

Epilepsy is a disease of the central nervous system. A seizure is an abnormal electrical activity in the brain of epileptic patients. One-third of children with autism spectrum disorder have seizures [1]. Seizure reduces the quality of life due to sudden loss of awareness or consciousness during various events. Anti-epileptic Drugs (AED) can be used for the treatment of epilepsy. But, according to the world health organization

R. Jana (✉) · S. Bhattacharyya
RCC Institute of Information Technology, Kolkata, India
e-mail: ranjan.rcciit@gmail.com

S. Bhattacharyya
e-mail: dr.siddhartha.bhattacharyya@gmail.com

S. Das
Indian Statistical Institute, Kolkata, India
e-mail: swagatam.das@isical.ac.in

© Springer Nature Singapore Pte Ltd. 2020
S. Bhattacharyya et al. (eds.), *Intelligence Enabled Research*,
Advances in Intelligent Systems and Computing 1109,
https://doi.org/10.1007/978-981-15-2021-1_7

(WHO), AED does not work effectively to reduce the seizure for 30% of the epileptic patients [2]. Electroencephalogram (EEG) is a type of electrical recording of signals in the brain activities by placing metal electrodes on the scalp. EEG signals have been often used as diagnostic tools for detecting upcoming seizure [3]. The states of the brain signal of epilepsy patients are generally classified into four categories [4]. Ictal is the state during a seizure event whereas preictal and postictal are other states before and after seizure events. Interictal is the normal state of the brain. The patterns of the brain EEG signals are different for four states as shown in Fig. 1. Application of AED on epilepsy patients for preventing seizures has some side effects and also it does not always effectively work for some patients. So, a correct prediction of upcoming seizure events can improve the quality of life by alarming the epilepsy patients. The brain EEG signal patterns are different for interictal, preictal, postitcal, and ictal states [4]. Since the interictal state is the normal state and preictal state being the state before seizure event, so identifying the preictal state is sufficient to predict the upcoming seizure event. Hence, the classification of interictal and preictal states is the main task for developing an upcoming seizure prediction system [5]. The brain EEG signal patterns are unparalleled for different epileptic patients [6]. Hence, a patient-dependent epileptic seizure prediction is generally proposed to achieve good performance. Furthermore, the seizure prediction is a challenging due to (i) limited EEG recording of preictal and ictal states; (ii) presence of noise in the EEG recording; (iii) and the difference in the EEG signal patterns for the same patient.

This paper is structured as follows: Section 2 describes related works. Section 3 shows the implementation details of the seizure prediction system. The experimental results and comparison study of the related works are shown in Sect. 4. Finally, conclusions are drawn in Sect. 5.

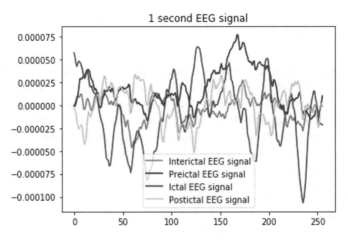

Fig. 1 Sample EEG signals of four states

2 Related Works

Epileptic seizure prediction by analysis of scalp EEG signal was explored for more than 20 years ago. Park et al. implemented a patient-dependent seizure prediction system by classifying the spectral power features of EEG signals [7]. They achieved 94.4% classification accuracy with a false prediction rate (FPR) of 0.2/h for 18 patients in the Freiburg EEG dataset. Williamson et al. extracted spatial and temporal features from EEG signals to develop a patient-dependent seizure prediction system [8]. They attained 85% classification accuracy with a FPR of 0.03/h for 19 patients in the Freiburg EEG dataset. Xian et al. extracted the fuzzy entropy features from the EEG signals and used the support vector machine to classify the features to develop the seizure prediction system [9]. They achieved an accuracy of 98.31%, sensitivity of 98.27% and specificity of 98.36% for the CHB-MIT dataset. Elgohary et al. extracted the features of EEG signals by counting the number of zero-crossings of wavelet detail coefficients [4]. They used a binary classifier that was used to classify the preictal and interictal states to implement the seizure prediction system. They attained 94% classification accuracy with a sensitivity of 96% for 8 patients in the Freiburg EEG dataset. Alotaiby et al. proposed a patient-dependent seizure prediction system using linear discriminant analysis (LDA) of the spatial patterns of EEG signals [10]. They achieved up to 89% sensitivity with a FPR of 0.39/h. Kitano et al. transformed the EEG signals into discrete wavelet forms to develop a patient-dependent seizure prediction method using Self Organizing Maps (SOM) [11]. They trained the system using only 20 min of EEG signals and achieved sensitivity up to 98%, specificity up to 88%, and accuracy up to 91% by using EEG signals of 9 epileptic patients from the CHB-MIT standard EEG dataset. Khan et al. also converted the EEG signals of 22 channels into a continuous wavelet form to construct a three-dimensional tensor (times, scales, and channels) of wavelet coefficients and used a convolutional neural network (CNN) to develop the seizure prediction system [12]. The system provided sensitivity up to 87.8% with a FPR of 0.142/h for the CHB-MIT standard EEG dataset. Truong et al. applied the short time Fourier transformation on raw EEG signals to extract the frequency and time domain features [13]. Then, they used CNN on the transformed features to develop a patient-dependent seizure prediction system. The prediction system provided sensitivity up to 81.4% and a FPR of 0.16/h with the CHB-MIT standard EEG dataset. Most of the researchers have used handcrafted feature extraction or any other transformation of raw EEG signals for predicting seizure events. In this paper, a novel technique is designed for epileptic seizure prediction from raw EEG signals of an epileptic patient. An image of the signal is constructed from raw EEG signals for analysis of signal patterns similar to the visual analysis of raw EEG signals carried out by medical practitioners. Then, the images of the signal are classified into interictal or preictal states using a CNN for predicting the upcoming seizure event. The proposed seizure prediction system shows promising performance.

3   Implementation

In this paper, a CNN model is proposed to predict the upcoming seizure event of an epilepsy patient. The seizure prediction method is framed to classify interictal and preictal states of the brain by analysis of the EEG signals. In traditional pattern recognition, the handcrafted features are usually extracted for any classification problem. The handcrafted features are extremely problem-dependent. Moreover, one expert is always required to select the good features for a specific problem. Lecun et al. discovered in 1995 that automatic feature extraction and classification from images can be done using CNN [14]. They proved that the accuracy using CNN is better compared to the accuracy provided using handcrafted feature extraction and classification for any classification problem. Most of the researchers have extracted the features or converted EEG signals into different forms before applying CNN for predicting the seizure of an epileptic patient [12, 13]. In real life, a physician analyzes a seizure event by observing the EEG signal patterns without any conversion of the signal. Thus, the authors are motivated to design a seizure prediction system by using the raw EEG signal. In this paper, the EEG signal is converted to form the image of the corresponding signal without any handcrafted features extraction or any transformation.

3.1   Database Used

In this paper, CHB-MIT EEG dataset [15] is used. This dataset was first used by Ali Shoeb in 2009 for his research work [16]. This EEG signal dataset is collected from pediatric epileptic patients of Children's Hospital, Boston. All the EEG recordings were captured at 256 samples per second with 16 bits for the intensity of each sample. The international standard of the 10–20 system with 23 channels was used for placing the electrodes on the scalp during EEG signal recording. The dataset is divided into two types of recording. One is recording without seizure during the normal activity of the epileptic patient and the other is recording with seizure when at least one seizure event is present. For simplicity, only 5 patients are considered for the same characteristics of the recording as shown in Table 1. The characteristics

Table 1   Recording of EEG signals used

Patient id	Recording # used for training	Recording # used for sensitivity	Recording # used for specificity
CHB01	06, 07, 03, 26	03, 26	06, 07, 08, 09, 10 11, 12, 13
CHB02	02, 03, 16+, 19	16+, 19	02, 03, 04, 05, 06, 07, 08, 09
CHB03	06, 07, 04, 34	04, 34	06, 07, 08, 09, 11, 12, 13, 14
CHB05	02, 03, 16, 17	16, 17	02, 03, 04, 08, 09, 10, 11, 24
CHB08	03, 04, 05, 11	05, 11	03, 04, 15, 16, 17, 18, 19, 23

are: (i) the duration of each recording is one hour, (ii) all patients should have at least two recordings with seizure, and (iii) 22 channels are fixed for all recording. Only 2 recordings with seizure and 2 recordings without seizure are considered in the training phase. For the testing phase, only 2 recordings with seizures are used to calculate the sensitivity and 8 recordings without seizure are used to calculate the specificity of the prediction system. A total of 10 recordings for each patient are considered for testing the system as shown in Table 1. Here, the duration of 1 s EEG signal is considered as one sample. Only 10 min (600 samples) preictal EEG signal data from each recording with seizure is considered. So, a total of 600 preictal samples from 2 recordings of each patient are used for training and cross-validation. Several researchers have taken the interictal signal data from EEG recording before 20 min of a seizure event. Hence, only 5 min (300 samples) of interictal EEG signal data from each recording without a seizure is considered to maintain an equal number of interictal and preictal samples for unbiased training. So, a total of 600 interictal samples from 4 recordings of each patient are used for training and cross-validation.

3.2 Preprocessing

A 1-second duration of EEG signal is considered as one sample as shown in Fig. 2. There are 256 discrete signal intensity values in one sample since the sampling rate is 256 Hz. To reduce the training time of the proposed system, only 64 discrete intensity values are considered with a gap of 4 values. Then, 64 discrete intensity values are plotted for generating a 64×64 image signal as shown in Fig. 3a. For each channel, the maximum and minimum signal intensity values are calculated and the intensity values are mapped into the corresponding row number. For example, the intensity values of column number 20 are mapped into row number 43 as shown in Fig. 3a. However, the plotted image shows that the generated line of the signal is not connected. So, the Bresenham line drawing algorithm is used to draw the connected line as shown in Fig. 3b. Hence, the dimension of the generated image of the signal is 64×64.

3.3 Convolutional Neural Network

The problem of seizure prediction is basically a classification of interictal and preictal states using the pattern of the EEG signals. CNN is capable to extract the features from images automatically and can classify the images efficiently based on the extracted features [14]. So, CNN is used here to implement an efficient seizure prediction system. In the begining, CNN uses a series of convolution and pooling operations for feature extraction. Finally, CNN uses one or more fully connected layers for the classification of the extracted features. Here, 64×64 images of 22 channels are considered for the EEG signal as one sample. So, the dimensions of the input layer of

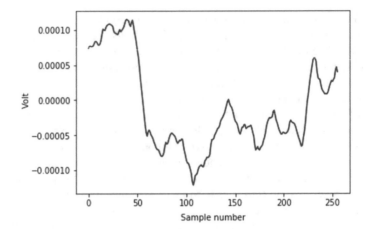

Fig. 2 A sample of 1 s EEG signal

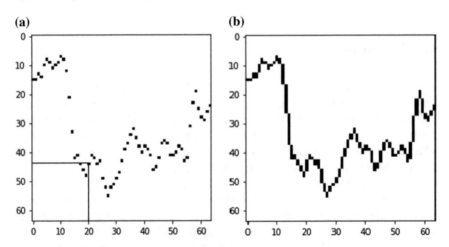

Fig. 3 **a** 1 s EEG signal plotting. **b** Using Bresenham line drawing algorithm

CNN are $64 \times 64 \times 22$. The CNN output layer comprises two nodes corresponding to two states: preictal and interictal. The proposed CNN architecture is shown in Fig. 4. Total 50 number of epochs is considered for the purpose of training and validation checking of the prediction system. The system is saturated after 40 epochs as shown in Fig. 5.

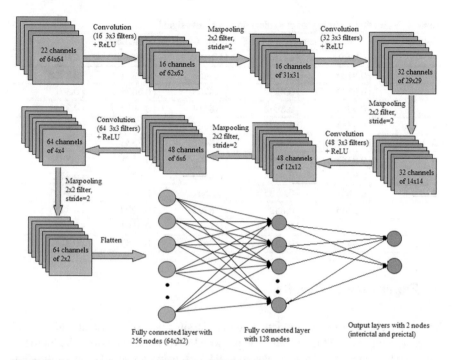

Fig. 4 Architecture of proposed convolutional neural network

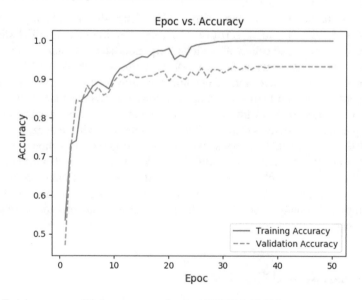

Fig. 5 Training versus validation accuracy of patient CHB01 in Fold-1

Table 2 Validation accuracy for classification of 1 s EEG signal

Fold	Validation accuracy using CNN (64×64 for 1 s signal data)				
	CHB01	CHB02	CHB03	CHB05	CHB08
1	0.9125	0.83337	0.8458	0.6833	0.8708
2	0.9333	0.8292	0.8708	0.6625	0.8917
3	0.9542	0.8042	0.8417	0.6917	0.8542
4	0.9583	0.8125	0.8667	0.6833	0.8792
5	0.9667	0.8500	0.8625	0.7333	0.9042
6	0.9333	0.8333	0.8667	0.7125	0.8833
7	0.9375	0.8458	0.8500	0.6917	0.8792
8	0.9667	0.8125	0.8667	0.6667	0.9042
9	0.9333	0.8500	0.8792	0.7000	0.8875
10	0.9375	0.8583	0.8708	0.7583	0.8875
Mean	0.9433	0.8329	0.8621	0.6983	0.8842

4 Experimental Results

The main aim of seizure prediction is the classification of interictal and preictal states by analysis of EEG signals. The proposed CNN model classifies the interictal and preictal states with an accuracy up to 94.33% using a tenfold cross- validation technique as shown in Table 2.

Medical practitioners use long duration of EEG signals for analyzing the signal patterns to predict upcoming seizures. Here, the duration of a 1 s EEG signal is considered for the classification of interictal and preictal states, which is not sufficient to predict upcoming seizures. So, duration of 1 min (60 samples) EEG signal is considered to take the decision of upcoming seizure prediction more accurately. The majority voting for pooled 60 samples is considered as the state of the patient. Recordings without seizures are considered to evaluate the specificity of the proposed seizure prediction system as shown in Table 3. Recordings with seizures are used to evaluate the sensitivity of the proposed seizure prediction system as shown in Table 3. Four time slots (0 to 5 min, 5 to 10 min, 10 to 15 min, and 15 to 20 min) before the seizure event are used to check the sensitivity for different slots. Finally, the proposed method is compared with the state-of-the-art prediction systems as shown in Table 4. The comparative analysis shows that the proposed methods provide a promising performance.

5 Discussions and Conclusion

This paper proposed a patient-dependent seizure prediction system to reduce the life risk of epileptic patients. The proposed system predicts the upcoming seizure event to inform the epileptic patients or their well-wishers to avoid unexpected death during unconsciousness. CNN is used for automatic feature extraction from raw EEG

Table 3 Specificity and sensitivity for 1 min pooled EEG signal

Patient id	Specificity (%)	Sensitivity in different time slot			
		0–5 min (%)	5–10 min (%)	10–15 min (%)	15–20 min (%)
chb01	100	100	100	100	90
chb02	91.12	100	100	90	60
chb03	94.79	100	100	100	80
chb05	87.71	100	80	50	40
chb08	98.33	100	100	100	70
Mean	94.39	100	96	88	68

Table 4 Comparison of performances of proposed system with others

Reference	Preprocessing /Feature extraction	Learning technique	Specificity	Sensitivity	Accuracy
Kitano et al. [11]	Discrete wavelet transform	SOM	Up to 88%	Up to 98%	Up to 91%
Khan et al. [12]	Continuous Wavelet transform	CNN	0.142 FP/h	87.8%	
Truong et al. [13]	Short time fourier transform	CNN	0.16 FP/h	81.2%	
Proposed method	EEG signal to image construction.	CNN	Up to 100% Average: 94.39% 0.0561 FP/h	100% for slot 0 to 5 min. 96% for slot 5 to 10 min. 88% for slot 10 to 15 min. 68% for slot 15 to 20 min.	Up to 94.33%

signals and classification of interictal and preictal states to decide on the upcoming seizure event. The proposed prediction system offers a promising performance. The comparative analysis shows that the proposed system outperforms the state-of-the-art methods for prediction of upcoming seizures. The seizure prediction is a challenging task due to (i) non-availability sufficient amount of EEG recording of preictal and ictal states; (ii) presence of noise during EEG recording; and (iii) the difference in the signal patterns for the same patients. Another challenging task is to design a patient-independent seizure prediction system in spite of having different patterns of each patient in their EEG signal. The authors are currently involved to solve the limitations of the proposed system.

References

1. Epilepsy Foundations, http://www.epilepsy.com
2. C.L. Deckers, P. Genton, G.J. Sills, D. Schmidt, Current limitations of antiepileptic drug therapy: a conference review. Epilepsy Res. **53**, 1–17 (2013)
3. M.Z. Parvez, M. Paul, Epileptic seizure detection by analyzing EEG signals using different transformation techniques. Neurocomputing **145**, 190–200 (2014)
4. S. Elgohary, S. Eldawlatly, M.I. Khalil, Epileptic seizure prediction using zero-crossings analysis of EEG wavelet detail coefficients, in *IEEE Conference on Computational Intelligence in Bioinformatics and Computational Biology*, pp. 1–6 (2016)
5. A.V. Esbroeck, L. Smith, Z. Syed, S. Singh, Z. Karam, Multi-task seizure detection: addressing intra-patient variation in seizure morphologies. Mach. Learn. **102**(3), 309–321 (2016)
6. J. Liang, R. Lu, C. Zhang, F. Wang, Predicting seizures from electroencephalography recordings: a knowledge transfer strategy, in *IEEE International Conference on Healthcare Informatics*, pp. 184–191 (2016)
7. Y. Park, L. Luo, K. Parhi, T. Netoff, Seizure prediction with spectral power of EEG using cost-sensitive support vector machines. Epilepsia **52**(10), 1761–1770 (2011)
8. J.R. Williamson, D.W. Bliss, D.W. Browne, J.T. Narayanan, Seizure prediction using EEG spatiotemporal correlation structure. Epilepsy Behav. **25**(2), 230–238 (2012)
9. J. Xiang, C. Li, H. Li, R. Cao, B. Wang, X. Han, J. Chen, The detection of epileptic seizure signals based on fuzzy entropy. J. Neurosci. Methods **243**, 18–25 (2015)
10. T.N. Alotaiby, S.A. Alshebeili, F.M. Alotaibi, S.R. Alrshoud, Epileptic seizure prediction using CSP and LDA for scalp EEG signals. Comput. Intell. Neurosci. **2017**, 1–11 (2017)
11. L.A.S. Kitano, M.A.A. Sousa, S.D. Santos, R. Pires, S. Thome-Souza, A.B. Campo, Epileptic seizure prediction from EEG signals using unsupervised learning and a polling-based decision process, pp. 117–126 (2018)
12. H. Khan, L. Marcuse, M. Fields, K. Swann, B. Yener, Focal onset seizure prediction using convolutional networks. IEEE Trans. Biomed. Eng. **65**(9), 2109–2118 (2018)
13. N.D. Truong, A.D. Nguyen, L. Kuhlmann, M.R. Bonyadi, J. Yang, S. Ippolito, O. Kavehei, Convolutional neural networks for seizure prediction using intracranial and scalp electroencephalogram. Neural Netw. **105**, 104–111 (2018)
14. Y. Lecun, Y. Bengio, Convolutional networks for images, speech, and time-series (1995)
15. CHB-MIT Scalp EEG Database, https://physionet.org/content/chbmit/1.0.0/
16. A. Shoeb, Application of machine learning to epileptic seizure onset detection and treatment, Ph.D. Thesis, Massachusetts institute of technology (2009)

A Deep Learning Approach for the Classification of Rice Leaf Diseases

Shreyasi Bhattacharya, Anirban Mukherjee and Santanu Phadikar

Abstract The fast and appropriate analysis and recognition of plant diseases can control the growth of diseases on various crops towards improving the quality and productivity of crops. The automatic system can perform disease recognition at minimum cost and error without the farm specialist's interpretation. It is very difficult to manually identify appropriate properties for distinguishing different kinds of crop diseases by using image processing and machine learning methods. In this study, we have developed a convolutional neural network (CNN) framework, a deep learning approach for automatically classifying three kinds of rice leaf diseases such as bacterial blight, blast, and brown mark. In the first phase, the developed system distinguished healthy and diseased leaves from a set of 1500 rice leaves. In the second phase, the three kinds of diseases have been categorized from a dataset containing 500 images of each of the three kinds of diseased rice leaves. The CNN model automatically learned required properties from raw images to differentiate the healthy and diseased rice leaves with 94% accuracy and then categorized different kinds of diseased rice leaves with 78.44% accuracy.

Keywords Rice leaf diseases · Rice leaf blast · Brown mark disease on rice leaf · Deep learning · Convolutional neural network

S. Bhattacharya (✉)
Department of IT, Maulana Abul Kalam Azad University of Technology,
Nadia 741257, West Bengal, India
e-mail: shreyasimakaut@gmail.com

A. Mukherjee
Department of Information Technology, RCC Institute of Information Technology,
Kolkata 700015, West Bengal, India
e-mail: anirbanm.rcciit@gmail.com

S. Phadikar
Department of CSE, Maulana Abul Kalam Azad University of Technology,
Nadia 741257, West Bengal, India
e-mail: sphadikar@yahoo.com

© Springer Nature Singapore Pte Ltd. 2020
S. Bhattacharyya et al. (eds.), *Intelligence Enabled Research*,
Advances in Intelligent Systems and Computing 1109,
https://doi.org/10.1007/978-981-15-2021-1_8

1   Introduction

In an agricultural country, essential daily food rice can be damaged due to various kinds of plant diseases apart from varying atmospheric conditions [8, 9, 12]. Rice diseases are mainly manifested by the effect of bacteria, fungus, and viruses that cause depletion of rice yield and thereby significantly affect the commercial health of the country. In agriculture, identification of disease crops and fruits is very essential to control the crop yield and quality.

Many researchers have proposed various automatic methods for the identification and classification of diseases of various crops.

The author of the paper [9] used hue saturation model (HIS), entropy-based bi-level thresholding procedure, 8-connectivity procedure, and self-organizing map (SOM) neural network in order to classify only two types of crucial rice diseases. The support vector machine (SVM) by choosing radial underlying kernel operation was applied to categorize the three types of rice diseases in the paper [12]. In a paper work [10], Otsu's threshold technique in the hue plane, Bayes' classifier, and SVM were used. In the paper [11], the various kinds of rice diseases were categorized by using Fermi energy based segmentation, genetic algorithm, and a rule base classifier. The paper [2] followed K-averages grouping method, gray level co-occurrence matrix, and discrete wavelet transform techniques. In the paper [3], the four types of rice diseases were classified by applying two classifiers like Minimum Distance Classifier (MDC) and k-Nearest Neighbor Classifier (k-NN). The backpropagation neural network (BPNN) method was applied to categorize rice leaf brown mark infection and normal rice leaf in the paper [6]. In order to identify two kinds of cucumber leaf diseases, the CNN technique was used under the 4-fold cross-validation method in the paper [4]. The four kinds of apple leaf diseases were classified by using a developed CNN model based on AlexNet model in the paper [5]. The paper [7] recognized rice pests and diseases by applying CNN system and image processing techniques. The paper [1] distinguished five kinds of mango leaf diseases by using self-operating deep CNN structure containing three hidden layers. The paper [13] used deep CNN system by combining three pretrained convolutional neural network systems AlexNet, GoogleNet, and ResNet for recognizing and categorizing eight kinds of tomato leaf diseases from uninfected leaf and stochastic gradient descent (SGD) was used.

The image processing and machine learning methods including the deep CNN method can automatically distinguish and classify various kinds of diseases of plants by extracting many properties in depth, thereby minimizing human effort and time as compared with the visual recognition of the diseases [1]. In this paper, among the several kinds of rice diseases, three kinds of rice leaf diseases namely, bacterial blight, blast, and brown spot are considered for the classification of rice diseases by using deep CNN method which to the best of our knowledge has not been done previously.

2 Proposed Work

The designed system has been used to identify the diseased images and healthy images of rice leaves and then to categorize the three kinds of diseased images of rice leaves successively. Figure 1 shows the component diagram for the proposed method.

2.1 Images Collection

The healthy rice leaf samples have been collected from a rice field of Nadia district, West Bengal, and images have been captured by using Canon EOS 1300D cameras (EF-S 18–55 IS II lens) at 4608 × 3456 pixels. The bacterial blight, blast, and brown spot infected rice leaf images have been collected from the Internet. The samples of collected healthy and infected rice leaf images have been shown in Fig. 2.

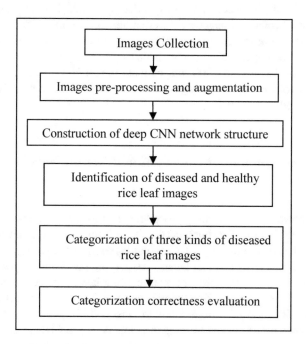

Fig. 1 Component diagram for the proposed method

(a)   **(b)**   **(c)**   **(d)**

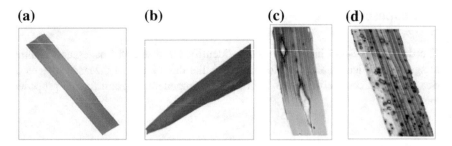

Fig. 2 Samples of collected rice leaf images: **a** normal rice leaf, **b** bacterial blight injured rice leaf, **c** blast-infected rice leaf image, **d** brown spot infected rice leaf image

2.2 Images Preprocessing and Augmentation

The black backgrounds of images have been removed as a preprocessing step. In order to train the deep convolutional neural network model, a large number of images are required. So, the dataset has been expanded by amplifying collected images in various ways: (i) rotating images by different angles like 30°, 45°, 60°, 90°, 180°, and 270°, (ii) moving images horizontally, and (iii) moving images vertically [13]. CNN can gain many kinds of properties from rotated images of different angles [6]. The final dataset has a total 2000 images that contain 500 images for every class of four classes of images. The images have been resized into 500 × 500 pixels to minimize time complications.

2.3 The Construction of Convolutional Neural Network Structure

A CNN is a powerful deep learning procedure and this process does not need manual parameter extraction. This contains one input layer, one output layer, and many hidden layers between those input and output layers. The hidden layer comprises a convolutional layer, cross-channel layer, max-pooling layer, and dropout layer. The convolutional layer, ReLU layer, and max-pooling layer together perform disease identification of crops by realizing many properties. The fully connected layer performs the classification of various kinds of classes in the output layer. Here, two phases have been performed by training the same designed CNN framework.

Phase 1: The identification of diseased images from healthy images of rice leaves

First, the diseased rice leaf images have been distinguished from the healthy rice leaf images by training the designed CNN model. In order to train the designed CNN model, the dataset of 1000 images has been partitioned into 70% train set and 30% test set.

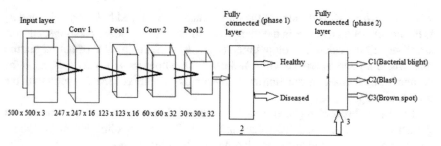

Fig. 3 Framework of designed CNN network

Table 1 The layers implementation of designed CNN network framework

Layers names	Operations/Layers performed	Filter sizes	Output feature sizes
Input			$500 \times 500 \times 3$
Conv1	Convolution	9×9	$247 \times 247 \times 16$
Maxpool1	Max-pooling	3×3	$123 \times 123 \times 16$
Conv2	Convolution	5×5	$60 \times 60 \times 32$
Maxpool2	Max-pooling	2×2	$30 \times 30 \times 32$
Output	Fully connected		$2 \times 1/3 \times 1$

Phase 2: The categorization of three kinds of rice leaf diseases

Second, the three kinds of diseased rice leaf images are categorized by training the designed model. The dataset of 1500 images has been split into train set and test set in the same way as in the first phase. The framework of designed CNN network for two phases is shown in Fig. 3.

The details of the layers of CNN network are given in Table 1.

Filter size: Width of the feature map × height of the feature map × no. of color channels (or feature maps of the previous layer).

Discussion: The CNN model has been designed using MATLAB software.

The images of size 500 × 500 have been given to the input layer and then proceeded through the designed CNN hidden layers. In order to perform convolution operation on the given images, 16, 32 filters of size 9 × 9, 5 × 5 successively with stride 2 and padding 0 have been utilized in the convolutional layers conv1 and conv2 of two hidden layers. The first and second convolution filter weights are (9, 9, 3, and 16) and (5, 5, 16, and 32), respectively. In the convolutional layer, the size of the generated output feature map is produced by,

$$w_2 = \left(\frac{w_1 - f + 2p}{s} + 1 \right), h_2 = \left(\frac{h_1 - f + 2p}{s} + 1 \right) \tag{1}$$

where $w_1 \times h_1 \times d_1$ is the size of the input image as w_1 is width, h_1 is height, and d_1 is the no. of color channels in the image. f is filter size, p is no. of paddings, and s is stride value. The calculated output map ($w_2 \times h_2 \times d_2$) is the size of output feature map where w_2 and h_2 are the width and height of output map, respectively, d_2 is the number of filters used in convolution operation. Bias value is 1 that is added with the produced output h_2 and w_2.

The ReLU layer changes all negative values into zeros (so that neurons cannot get activated) and all positive values into one. The max-pooling layers maxpool1 and maxpool2 are applied by employing 16, 32 filters of size 3×3, 2×2 successively with stride 2 and padding 0 to perform the max-pooling operation. The max-pooling layer finds the maximum value among the pixels in each partitioned non-overlapping block in the feature map that was produced from previous convolutional layer and generates the feature map of minimized coordinate size to diminish number of unnecessary parameters. After performing each max-pooling layer operation, 35% and 25% dropout has occurred in the dropout layers. The fully connected layer generates two-dimensional and three-dimensional linear vectors for phases 1 and 2, respectively, according to the number of classes. The test set images have been tested. In the implemented CNN network, some of the training options that are used are 10 epochs, minibatch size of 16, initial learning rate of 0.04, and momentum of 0.9.

2.4 Training–Progress Graphs of the Designed CNN Model

The network training–progress graphs for the two phases are shown in Figs. 4 and 5, respectively.

These progress graphs for the two phases signify that while the number of iterations for each epoch increases, the training accuracy increases smoothly or sometimes decreases and the loss decreases at the same time.

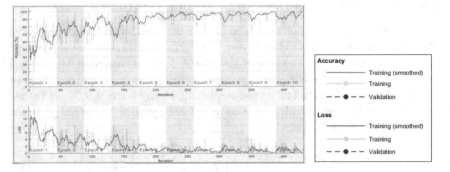

Fig. 4 Network training progress graph for phase 1

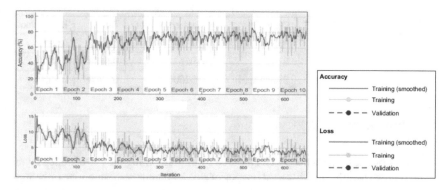

Fig. 5 Network training progress graph for phase 2

3 Experimental Results

The designed CNN network has been trained with the train set of rice leaf images at 10 and 14 epochs. Thereafter, the test set images have been tested and this designed model can give much accuracy for phase 1 and phase 2 at 10 and 14 epochs both. Categorization correctness can be calculated as follows

$$\text{Categorization correctness} = \frac{\text{Number of truely categorized images}}{\text{Total number of images}} \quad (2)$$

Table 2 shows the system's categorization accuracies for datasets of two modules under training options.

Confusion matrix: The confusion matrixes for both phases are shown in Tables 3 and 4.

Table 2 System's categorization accuracies under experimental conditions

Phases	Number of images	Number of epochs	Minibatch sizes	Training times	Classification accuracies
1	1000	10	16	21 min 23 s	**94%**
1	1000	14	16	78 min 35 s	95.67%
2	1500	10	16	32 min 04 s	**78.44%**
2	1500	14	16	45 min 24 s	73.56%

10 epoch results are highlighted using bold

Table 3 Confusion matrix of phase 1 at 10 epochs

Class	Healthy	Diseased
Healthy	0.9400	0.0600
Diseased	0.0600	0.9400

Table 4 Confusion matrix of phase 2 at 10 epochs

Class	Bacterial blight	Blast	Brown spot
Bacterial blight	0.6333	0.1467	0.0022
Blast	0.0267	0.8133	0.1600
Brown spot	0.0200	0.1800	0.8000

4 Conclusion

This work has designed a deep CNN method by implementing only two hidden layers to distinguish healthy and diseased images of rice leaf; and then categorize only three kinds of rice leaf diseases with reasonable perfection and good training performance but with some error. It is observed that varying different options of CNN network may not improve the disease identification and categorization accuracy of rice leaves in the case of three or more hidden layers. In future, a larger dataset of best quality natural images of different diseased parts of rice plants will be collected and the CNN framework will be modified by implementing three or more hidden layers with necessary fine-tuning of parameters. The other types of CNN models like AlexNet, GoogleNet, ResNet will be tried for distinguishing and categorizing the rice diseases.

References

1. S. Arivazhagan, S. Vineth Ligi, Mango Leaf Diseases Identification Using Convolutional Neural Network. Int. J. Pure Appl. Math. **120**(6), 11067–11079 (2018)
2. R. Deshmukh, M. Deshmukh, Detection of paddy leaf diseases. in *InInternational Conference on Advances in Science and Technology 2015* (ICAST 2015)
3. A.A. Joshi, B.D. Jadhav, Monitoring and controlling rice diseases using image processing techniques. in *2016 International Conference on Computing, Analytics and Security Trends (CAST)* (IEEE, 2016)
4. Y. Kawasaki et al., Basic study of automated diagnosis of viral plant diseases using convolutional neural networks. in *International Symposium on Visual Computing* (Springer, Cham, 2015)
5. B. Liu et al., Identification of apple leaf diseases based on deep convolutional neural networks. Symmetry **10**(1), 11 (2017)
6. L. Liu, Z. Guomin, Extraction of the rice leaf disease image based on BP neural network. in *2009 International Conference on Computational Intelligence and Software Engineering* (IEEE, 2009)
7. E.L. Mique Jr, T.D. Palaoag, Rice pest and disease detection using convolutional neural network. in *Proceedings of the 2018 International Conference on Information Science and System* (ACM, 2018)
8. M. Mukherjee, T. Pal, D. Samanta, Damaged paddy leaf detection using image processing. J. Glob. Res. Comput. Sci. **3**(10), 07–10 (2012)
9. S. Phadikar, J. Sil, Rice disease identification using pattern recognition techniques. in *2008 11th International Conference on Computer and Information Technology* (IEEE, 2008)
10. S. Phadikar, J. Sil, A. Kumar Das, Classification of rice leaf diseases based on morphological changes. Int. J. Inform. Electron. Eng. **2**(3), 460–463 (2012)

11. S. Phadikar, J. Sil, A. Kumar Das, Rice diseases classification using feature selection and rule generation techniques. Comput. Electron. Agric **90**, 76–85 (2013)
12. Q. Yao et al., Application of support vector machine for detecting rice diseases using shape and color texture features. in *2009 International Conference on Engineering Computation* (IEEE, 2009)
13. K. Zhang et al., Can deep learning identify tomato leaf disease? Adv. Multimedia (2018)

Identification of Seven Low-Resource North-Eastern Languages: An Experimental Study

Joyanta Basu⊙ and Swanirbhar Majumder⊙

Abstract This paper describes an experimental study on the identification of seven north-eastern (NE) low-resource languages (LRL) of India namely, Assamese, Bengali, Hindi, Manipuri, Mizo, Nagamese, and Nepali using 55-dimensional hybrid features (HF) and hierarchical technique. However, the development of language identification (LID) systems could be leveraged only with the availability of specially curated speech data in LRL and it is a really challenging task to build a system on such under-resourced languages. We have collected around 42 h of speech data (including 35 h for training and 7 h for testing) for analysis on the above-said seven LRL. The process of designing speech database in LRL has been generic enough to be used for other languages as well. We have compared our proposed methodology with baseline system on collected speech data. From the experimental study, it has been observed that our proposed system is outperformed over baseline system and results are encouraging for researchers in low-resource languages. This initial study unveils the importance of HF for NE-LRL.

Keywords Language identification (LID) · North-Eastern (NE) Low-resource language (LRL) · Hybrid features (HF)

1 Introduction

LID is the problem of identifying the language being spoken from a sample of speech. It is also called as spoken language recognition (SLR). Automatic LID is an integral part of multilingual conversational systems, multilingual speech and speaker

J. Basu (✉)
CDAC, Kolkata, Salt Lake, Sector - V, Kolkata 700091, India
e-mail: joyanta.basu@cdac.in

S. Majumder
Department of Information Technology, Tripura University,
Suryamaninagar 799022, Tripura, India
e-mail: swanirbhar@ieee.org

© Springer Nature Singapore Pte Ltd. 2020
S. Bhattacharyya et al. (eds.), *Intelligence Enabled Research*,
Advances in Intelligent Systems and Computing 1109,
https://doi.org/10.1007/978-981-15-2021-1_9

recognition, spoken language translation, and spoken document retrieval. In a multilingual country like India, automatic LID systems have a special significance. With the mass usage of smartphones and computers in our day-to-day activities, the need for such systems to break the language barrier can easily be understood. For example, an interactive voice response (IVR) system catering to customers speaking different languages can use a front-end LID or SLR module to route incoming calls [1]. Online audio documents can be efficiently indexed and searched based on the languages being spoken. Law enforcement organizations may find this technology useful for surveillance and security applications in the globalized world. Vowel duration, formants, and other features like Mel-Frequency Cepstral Coefficients (MFCC) [2] and Linear Predictive Cepstral Coefficients (LPCC) [3] play a very important role in characterizing the speech signal as well as identifyingany individual's speech elements or phoneme. Formants represent the vocal track resonance frequency. Formants have been long regarded as one of the most compact and descriptively powerful parameter sets for voiced speech sound, with important correlates in both auditory-perceptual and articulatory domains [4]. It also a cue to the intelligibility of human speech. From previous study, it was examined that both formants and prosodic feature vectors are useful for language and speaker identification. It was also found that formant features were generally superior to prosodic features. The formant vector-based LID system used k-means clustering [5].

Based on the availability of resources, languages can be categorized as well-resourced and under-resourced languages. The term under-resource refers to languages with one or more of the following aspects: lack of a unique writing system or stable orthography, lack of linguistic expertise, lack of electronic resources for speech, and language data [6]. Modern India, as per the 1991 census, has more than 1576 mother tongues, genetically belonging to five different language families. They are further rationalized into 216 mother tongues, and grouped under 114 major languages [7]. The year 2001 census identified 122 major languages in India, out of which 29 languages have more than a million native speakers, while other 1599 languages are spoken by smaller societies, local groups, and tribes. As per Eighth Schedule, there are 22 official languages in India [8]. Though having such a rich language heritage and enormous diversity, all the Indian languages are under-resourced in terms of digitally available language resources. Designing an effective framework to create and preserve digitally available resources for under-resourced languages is very important in the Indian scenario.

2 Purpose of Work

In this paper, we worked on LID of seven NE-LRL namely, Assamese (AS), Bengali (BN), Hindi (HN), Manipuri (MA), Mizo (MI), Nagamese (NA), and Nepali (NP). Though are more than 220 languages as per the 1971 census [9]. This is rarely being studied earlier together for LID of seven NE-LRL perspectives. In this work, therefore our aim is to identify languages from audio data in natural speaking scenarios and investigate the same in detail to figure out identification performance. The novelty of this paper is to work with NE-LRL and develop a suitable model to identify the unknown languages using hierarchical techniques and hybrid features (HF).

3   About Low-Resource Languages of North-East India

NE languages are spoken in eight states of north-eastern part of India and they belong mainly to three language families, namely Indo-Aryan, Sino-Tibetan, and Austro-Asiatic. We have selected seven NE-LRL for the study.

Table 1 shows the basic information as well as comparison of the above-said seven NE-LRL.

Table 1 Information of low-resource Indian languages of north-eastern states

#	Language family	Language name	Location	Population	No. of vowels	No. of consonants	Tonal/non-tonal
1	IEF	AS	Assam	12,800,000 in India (2001 census)	8	23	Non-tonal
2	IEF	BN	Assam, Tripura, West Bengal	210 million speakers in world (2011 census)	7 (Oral) and 7 (Nasal)	33	Non-tonal
3	IEF	HN	Throughout north India and north-eastern states	180,000,000 in India (1991 UBS)	13	34	Non-tonal
4	STF	MA (Meitei)	Mainly Manipur, some districts of Nagaland, Assam, Mizoram	1,470,000 in India (2001 census)	6	24	Tonal (Two tones)
5	STF	MI	Mizoram	674,756 speakers in India (2001 census)	6	33	Tonal (Four tones)
6	IEF	NA	Nagaland, Arunachal Pradesh	1,980,602 per the 2011 census	6	26	Non-tonal
7	IEF	NP	Sikkim	25 million (2001–2011 censuses)	11	29	Non-tonal

4 Data Preparation

4.1 Experimental Data Set

Preparation of experimental data set is really a challenging task for LRL. Some of the languages that do not have any written scripts, use Latin script like Nagamese and Mizo. We have created text materials in each of the seven languages. Text materials include digits, numbers, and paragraphs on different topics for each language. We have also given English paragraphs for manual translation in native language for recording. Recording data set also includes free speech on any selected topics, free speech on different scenery of pictures, and finally conversation between two speakers. The materials are read out by 70 native speakers (10 per language) including 30 female speakers from north-east India within the age group 20–60 years.

4.2 Data Collection

All the speech data has been collected in room environment to reduce external noise, using a good quality headset with standard noise cancelation facilities. To minimize the lack of fluency in reading, informants are instructed to rehearse at least once before the final recording. We have collected nearly 6 h of speech data in each language including two repetitions. Speech data is recorded with 16-bit mono Microsoft wave PCM format and 22,050 Hz sampling rate using the Praat software (an open-source speech-editing software) [10]. So, for the analysis purpose, around 42 hours of speech data was collected, in which we have used 35 h' data as training for the system and 7 h of data is used to test the performance of the proposed system.

4.3 Data Transcription

All the recorded speech data are transcribed using the Praat software. All speech data are segmented automatically to identify the voice and unvoiced region using Praat scripts. After automatic segmentation human transcribers carefully cross-check each sound file, adjust marked start–end boundaries for different events like environmental noise (Noise), silence (SIL), voiced (V), and unvoiced (UV) section.

5 Proposed Architecture and Methodology

Figure 1 shows the proposed general block diagram of the language identification system for low-resource north-east Indian languages.

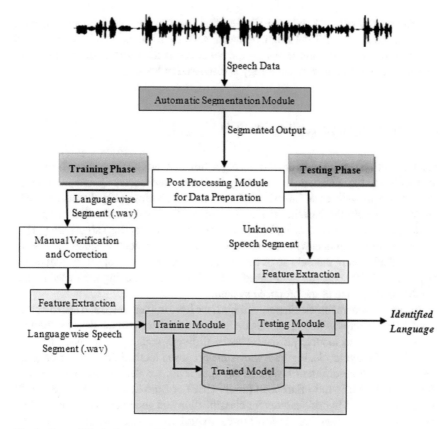

Fig. 1 General block diagram of the proposed system

Speech signal needs to be segmented automatically and followed by manual verification and correction of the training data. After extraction of features entered into two important modules, one is training and the other is the testing module. Description of modules is given below.

5.1 Feature Extraction

This is an important task with respect to system accuracy. Collected speech data should be carefully analyzed and preprocessed (noise reduction, channel compensation, DC normalization, voice activity detection, etc.) automatically such that important features remain intact. Features like MFCC with delta and double delta, and Partial Correlation (PARCOR) coefficients [11] are used with addition of formant frequencies of voiced region. Here in this proposed system, we have used hybrid features (HF), i.e., 13 MFCC + 13 Delta + 13 Double Delta + 13 PARCOR +

3 Formants (namely, F1, F2, and F3, i.e., first formant, second formant, and third formant, respectively). Total length of the HF vector is 55. In this work, we have used feature vectors on different levels for the benefit of system performance and due to LRL challenge. Detailed methodology is discussed in Sect. 6.

5.2 Training and Testing Module

Selection of system training or learning algorithm depends on the nature of collected data and target application. In this study, we have used Vector Quantization (VQ) [12] -based as well as a Gaussian mixture model (GMM) [13] -based language modeling to train the baseline model. Maximum likelihood estimation of speaker model parameters is done using EM (Expectation–Maximization) algorithm [14].

The testing set of data will be used for evaluating the system. Besides, the system shall be deployed in the field so as to evaluate the performance of the system under practical real-life scenarios. VQ-based and GMM-based matching were used to find out the performance of the proposed system.

As per the general block diagram of Fig. 1, we have implemented and tested hybrid feature (HF) vector with VQ and GMM. As we do not have much data for training the language-wise model, we have chosen different techniques to make features hybrid at different levels. Figure 2 shows the proposed methodology of our ongoing work. It has three different levels. In the first level, we determined the language family whether it is Indo-European family (IEF) or Sino-Tibetan family (STF). In level two, we used hybrid features for classification and generate score of individual languages of unknown data. In level three, hybrid score is calculated and decision is taken to identify the unknown low-resource language. In level two, we have used HF for classification but in level one, we used only MFCC, Delta, Double Delta, and PARCOR coefficient as a feature set. In both of the levels, we used VQ and GMM as classifiers for analysis.

6 Results and Discussion

In this section, we will discuss first our baseline system performance and then we will discuss the proposed system performance on top of the baseline system.

6.1 Performance of Baseline System

In the baseline system, we have extracted MFCC, Delta, Double Delta, and PARCOR coefficients features (i.e., vector size is 52) and develop language-wise VQ and GMM models separately. In the experiments, we do not consideration language family and

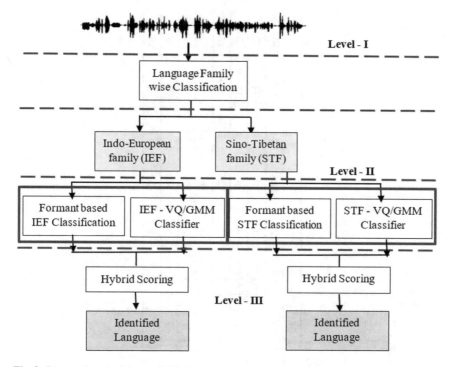

Fig. 2 Proposed methodology of identification of low-resource North-Indian languages

formants' feature characteristics of NE languages. From Table 2, it has been observed the confusion matrix (CM) is in percentage using VQ and GMM classifiers. Using GMM, performance is improved than VQ. As an example, we can say Assamese language performance is confused with Bengali and Nagamese language mainly due to being in the same language family as well as similar language structures in nature. The same is true for Nepali and Hindi languages. Similar phenomena can be observed for Manipuri and Mizo language because both of them are STF. After analysis of Table 2, we can say that baseline system achieved around 55% and 72% accuracy in VQ- and GMM-based model, respectively, irrespective of any language.

6.2 Performance of Language Family Classification Module

Now, as per our proposition, we have measured the performance of Language family in level one. In this level we have used baseline methods to identify the language family i.e. we have extracted 52 vector feature coefficients and modeled with VQ and GMM. Table 3 shows the CM to identify the language family. It has been observed that GMM-based approach outperformed the VQ-based model. It achieved 84% accuracy for STF and 78% for IEF in GMM-based approach.

Table 2 CM using MFCC and PARCOR coefficients with VQ and GMM Classifiers

In %	VQ classifiers							GMM classifiers						
	AS	BN	HN	MA	MI	NA	NP	AS	BN	HN	MA	MI	NA	NP
AS	**39**	21	7	6	5	15	7	**65**	17	1	2	2	11	2
BN	15	**50**	4	9	2	14	6	11	**67**	3	3	4	10	2
HN	4	3	**52**	3	6	12	20	2	2	**72**	2	3	7	12
MA	1	13	2	**65**	14	4	1	1	1	1	**76**	18	2	1
MI	1	2	4	23	**63**	3	4	1	1	2	13	**78**	3	2
NA	5	12	9	7	3	**62**	2	3	9	5	4	3	**73**	3
NP	2	7	26	5	1	3	**56**	1	5	8	3	1	3	**79**

Table 3 CM of language family performance using HF with VQ and GMM classifier

In %	VQ classifier		GMM classifier	
	Indo-European	Sino-Tibetan	Indo-European	Sino-Tibetan
Indo-European	**65**	35	**78**	22
Sino-Tibetan	28	**72**	16	**84**

6.3 Performance Analysis of Proposed Methods

In this section, we analyzed level-wise classification as discussed in Sect. 6. We have extracted 52 vector feature coefficients as well as language-wise format frequencies, i.e., F1, F2, and F3 from the voiced region of the signal. So, feature vector length for training and testing is 55. This is called HF. Table 4 shows the performance of the proposed model using VQ- and GMM-based approach, respectively. This performance is after hybrid scoring techniques. From the tables it has been observed that the GMM-based model using MF outperforms the VQ-based approach and it also outperformed the baseline system as described above. Identification of two STF languages, i.e., Manipuri and Mizo are on the higher side using GMM, i.e., 90% and 89%, respectively. Within the IEF languages, Nepali language performs better (88% in GMM-based approach) than others in the same language family. Finally, from the proposed system we achieved 70% and 82% in VQ- and GMM-based approach, respectively, after using HF and language family recognition and hybrid scoring mechanism.

7 Conclusions

In this paper, we have reported a detailed study of seven NE-LRL of India based on language family as feature, as well as formant frequencies and MFCC, Delta, Double Delta, and PARCOR coefficients. An experimental study was also carried out to find out the role of HF in language discrimination. In the baseline system, we have achieved language recognition to be around 55%. It has been confused with other languages because of similar language family and structure like Assamese–Bengali–Nagamese. On introducing formants frequencies in feature set as well as language family-wise recognition on top, recognition rate gets increased for STF like Manipuri and Mizo to 90% and 89%, respectively, and also for Indo-European languages. However, it needs further investigation to improve the performance of both IEF as well as STF languages. This preliminary study may help researchers in the area of other language identification of low-resource north-eastern languages. However, there is scope of further study on tonal language features like fundamental frequency (F0) of STF languages to improve the performance. As in the NE, there are many dialectical variations; we need to consider dialectical variational features for more robust low-resource language identification.

Table 4 CM using HF with VQ and GMM classifiers with consideration of language family

In %	VQ classifiers							GMM classifiers						
	AS	BN	HN	MA	MI	NA	NP	AS	BN	HN	MA	MI	NA	NP
AS	**54**	15	5	5	4	10	7	**73**	13	1	2	1	8	2
BN	12	**63**	3	4	1	12	5	9	**76**	2	2	2	8	1
HN	2	2	**67**	4	3	7	15	2	2	**78**	2	3	5	8
MA	1	2	0	**88**	8	1	0	1	1	0	**90**	7	1	0
MI	1	1	1	11	**79**	3	4	0	0	1	7	**89**	2	1
NA	4	8	9	3	1	**74**	1	2	5	2	1	3	**86**	1
NP	2	6	19	4	1	3	**65**	0	2	6	1	0	3	**88**

Acknowledgements Authors would like to acknowledge the contribution of NERIST, Arunachal Pradesh for supporting speech data collection from native speakers of north-east Indian. The authors are thankful to CDAC, Kolkata, India for necessary financial and infrastructural support to carry out research activity.

References

1. Y.K. Muthusamy, E. Barnard, R.A. Cole, Reviewing automatic language identification. IEEE Signal Process. Mag. **11**(4), 33–41 (1994)
2. D. Jurafsky, J. Martin, *Speech and Language Processing: An Introduction to Natural Language Processing*, 2nd edn. (Prentice Hall, New Jersey, 2008)
3. L. Rabiner, B. Juang, *Fundamentals of Speech Recognition* (Prentice Hall, New Jersey, 1993)
4. A. Itchikawa, Y. Nakano, Nakata, Evaluation of various parameter sets in spoken digits recognition. IEEE Trans. Audio Electro-acoust **AU-21**(3) (1973)
5. J.T. Foil, Language identification using noisy speech, in *International Conference on Acoustics, Speech, and Signal Processing (ICASSP)* (1986), pp. 861–864
6. L. Besacier, E. Barnard, A. Karpov, T. Schultz, Automatic speech recognition for under-resourced languages: a survey. Speech Commun. **56**, 85–100 (2014)
7. http://mhrd.gov.in/language-education
8. Languages of India, https://en.wikipedia.org/wiki/Languages_of_India
9. J. Basu et al., Acoustic analysis of vowels in five low resource North East Indian languages of Nagaland, in *O-COCOSDA* (Seoul, Korea (South), 2017), pp. 145–150 (2017)
10. Praat Website (2016), http://www.fon.hum.uva.nl/praat/
11. T. Soni, J.R. Zeidler, W.H. Ku, Behavior of the partial correlation coefficients of a least squares lattice filter in the presence of a nonstationary chirp input. IEEE Trans. Signal Process. **43**(4), 852–863 (1995)
12. Y. Linde, A. Buzo, R.M. Gray, An algorithm for vector quantizer design. IEEE Trans. Commun. **28**(1), 84–95 (1980)
13. D.A. Reynolds, R.C. Rose, Robust text-independent speaker identification using Gaussian mixture speaker models. IEEE Trans. Speech Audio Process. 72–83 (1995)
14. A. Dempstar, N. Larid, D. Rubin, Maximum likelihood from incomplete data via the EM algorithm. R. Stat. Soc. B **39**, 1–38 (1977)

Implications of Nonlinear Control Over Traditional Control for Alleviating Power System Stability Problem

Asim Halder, Nitai Pal and Debasish Mondal

Abstract One of the major ways to improving the power system stability is the implication of highly effective technology in control system design. Linear controllers are usually designed through approximate linearization of a nonlinear system around a single operating condition that is not effective in severe contingencies and stressed operating conditions. In this regard, the design and application of nonlinear control strategy is the best-suited solution. The nonlinear controllers can be effectively designed through exact linearization of a nonlinear system and therefore the states of the system do not lose their originality. Moreover, nonlinear controllers are not only suitable for multiple operating points but also able to mitigate small as well as large disturbances. Based on the above requirements, authors in this research aim to investigate the dynamic instability problem of a power system and its alleviation through deployment of nonlinear controller.

Keywords Game theory $\cdot$ H_∞ controller $\cdot$ Linear matrix inequality $\cdot$ Nonlinear control $\cdot$ Conventional control $\cdot$ Static synchronous compensator $\cdot$ Zero dynamic design $\cdot$ Particle swarm optimization $\cdot$ Genetic algorithm

1 Contributions of the Research Undertaken

In this research work, various approaches—Zero Dynamic Design, Feedback Linearization, and Nash Equilibrium of Game Theory—are adopted to design nonlinear

A. Halder (✉)
Department of Applied Electronics and Instrumentation Engineering, HIT, Haldia, India
e-mail: asim_calcutta@yahoo.com

N. Pal
Department of Electrical Engineering, Indian Institute of Technology (ISM), Dhanbad, India
e-mail: nitai@iitism.ac.in

D. Mondal
Department of Electrical Engineering, RCC Institute of Information Technology, Kolkata, India
e-mail: mondald4791@gmail.com

© Springer Nature Singapore Pte Ltd. 2020
S. Bhattacharyya et al. (eds.), *Intelligence Enabled Research*,
Advances in Intelligent Systems and Computing 1109,
https://doi.org/10.1007/978-981-15-2021-1_10

controller [1, 2]. It has been well-reported in the literature that the Flexible Alternating Current Transmission Systems (FACTS) devices are very efficient and have the potential capability to solving power system problems [3, 4]. They have become the most important power system components in the present Smart Grid (SG) scenarios. Therefore, the design of nonlinear control scheme for FACTS devices, especially for Static Synchronous Compensator (STATCOM) is taken into consideration [5, 6]. The efficacy and performance of these proposed nonlinear controllers have been investigated in comparison to conventional linear controller for application in power systems. To the best of the author's knowledge, limited works have been carried out by researchers in these directions. To design nonlinear FACTs controller for power system stability improvement and its performance investigation with respect to the conventional linear controllers in the face of power system contingencies, the following approaches have been taken into consideration.

2   Deterministic Approach for the Design of Nonlinear Controller

2.1   Nonlinear Excitation Controller (NEC) via Zero Dynamics Design Approach

The classical third-order model of a Single Machine Infinite Bus (SMIB) system has been considered in this work (Fig. 1) for investigation of stability and the method of zero dynamic design [7] has been implemented. The SMIB system equations are

$$\dot{E}'_q = -\frac{1}{T'_d} E'_q + \frac{1}{T_{do}} \frac{x_d - x'_d}{x'_{d\Sigma}} V_s \cos \delta + \frac{1}{T_{do}} V_f \tag{1}$$

Fig. 1 Schematic of nonlinear excitation controller with SMIB study system

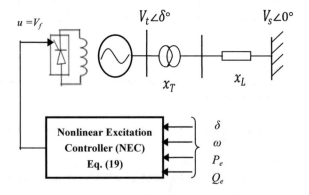

$$\dot{\omega} = \frac{\omega_0}{H} P_M - \frac{D}{H}(\omega - \omega_0) - \frac{\omega_0}{H} \frac{E_q' V_s}{x_{d\Sigma}'} \sin \delta \tag{2}$$

$$\dot{\delta} = (\omega - \omega_0) \tag{3}$$

Based on the zero dynamic design theory, a coordinate transformation need from original state "X" to the desired state "Z" is given by

$$\dot{Z}_1 = L_f f(X) = Z_2 \tag{4}$$

$$\dot{Z}_2 = L_f^2 h(\varphi^{-1}(Z)) + L_g L_f h(\varphi^{-1}(Z))u \tag{5}$$

$$\dot{Z}_3 = L_f \varphi_3(\varphi^{-1}(Z)) \tag{6}$$

where $Z = \begin{bmatrix} Z_1 & Z_2 & Z_3 \end{bmatrix}^T$; L_f and L_f^2 are first and second derivative along $f(X)$. Now, following the guidelines of zero dynamic design method the nonlinear control law has been derived as [8] (Fig. 2);

$$0 = L_f^2 h(\varphi^{-1}(Z)) + L_g L_f h(\varphi^{-1}(Z))u \tag{7}$$

$$u = -\frac{L_f^2 h(\varphi^{-1}(Z))}{L_g L_f h(\varphi^{-1}(Z))} = -\frac{L_f^2 h(X)}{L_g L_f h(X)} \tag{8}$$

Finally, the control law becomes

$$u = V_f = E_q - \frac{T_{do} E_q'}{P_e}\left(Q_e + \frac{V_s^2}{x_{d\Sigma}'}\right)\Delta\omega - \frac{T_{do} E_q' D}{H}\dot{\omega} \tag{9}$$

Fig. 2 STATCOM controller installed in SMIB system

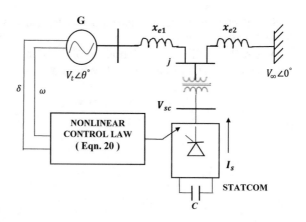

where $E_q = \frac{x_{d\Sigma}}{x'_{d\Sigma}} E'_q - (x - x'_d) \frac{V_s}{x'_{d\Sigma}} \cos \delta$; $Q_e = \frac{V_s E'_q}{x'_{d\Sigma}} \cos \delta - \frac{V_s^2}{x'_{d\Sigma}}$; and $P_e = \frac{V_s E'_q}{x'_{d\Sigma}} \sin \delta$.

2.2 Design of Nonlinear SATCOM Controller of a Power System

To develop the exact linearization based nonlinear control scheme, the governing dynamic model of an SMIB test system has been considered as [9].

$$\dot{\delta}(t) = \omega(t) - \omega_s \tag{10}$$

$$\dot{\omega}(t) = \frac{\omega_s}{2H} \left(P_M - \frac{D(\omega(t) - \omega_s)}{\omega_s} - P_e \right) \tag{11}$$

where all the parameters of (10), (11) have their usual meaning.

If the control input signal "u" is chosen to be "u" = $V_{sc.}$ Then (10), (11) can be represented by the following standard affine nonlinear form as

$$\dot{X}(t) = \tilde{f}(X) + \tilde{g}(X)u \tag{12}$$

$$\tilde{y}(t) = \tilde{h}(X) \tag{13}$$

where $X = [\omega \ \delta]^T$; $\tilde{f}(X) = \begin{bmatrix} \frac{\omega_s}{2H}(P_M - \frac{D}{\omega_s}(\omega - \omega_s)) \\ (\omega - \omega_s) \end{bmatrix}$; and $\tilde{g}(X) = \begin{bmatrix} -\frac{\omega_s}{2H} \frac{E'_q V_\infty \sin \delta}{G_1} \\ 0 \end{bmatrix}$.

The function $\tilde{h}(X)$ denotes the output vector. Now estimation of the control law is presented as follows.

A suitable coordinate transformation of the affine system (12), (13) reduces the system into normal form given by *Brunovsky* as

$$\dot{z}_1 = z_2 \tag{14}$$

$$\dot{z}_2 = \gamma \tag{15}$$

This is the exactly linearized system with "γ" designated as its "optimal control" input and state vectors are $z_1 = \delta - \delta_0$ and $z_2 = \omega - \omega_0$. Now (14) and (15) together can be expressed in standard state variable form as

$$\dot{Z} = AZ + B\gamma; \; Z(0) = Z_0 \tag{16}$$

The connection between the control input "u" and the optimal control input "γ" of the *Brunovsky* system (14), (15) can be obtained as

$$u = -\frac{\alpha(X)}{\beta(X)} + \frac{1}{\beta(X)}\gamma \tag{17}$$

where $\alpha(X) = L_f^2 h(X) = \left(\frac{\omega_s P_M}{2H} - \frac{D(\omega(t) - \omega_s)}{2H}\right)$ and $\beta(X) = L_g L_f h(X) = -\frac{\omega_s E_q' V_\infty \sin \delta}{2HG_1}$

Therefore, the control law following Eq. (17) gives

$$u = -\frac{\left(\frac{\omega_s P_M}{2H} - \frac{D(\omega(t) - \omega_s)}{2H}\right)}{-\frac{\omega_s E_q' V_\infty \sin \delta}{2HG_1}} + \frac{1}{-\frac{\omega_s E_q' V_\infty \sin \delta}{2HG_1}}\gamma \tag{18}$$

Now, to acquire the *control law* of optimality for the above exactly linearized system (14), (15) of the original nonlinear power system, the principle of *Linear Quadratic Regulator* (LQR) can be applied which is obtained as

$$\gamma = -K^* Z = -K_1^* z_1 - K_2^* z_2 \tag{19}$$

The gain matrix K of the state feedback controller is calculated as $K^* = [K_1 \; K_2] = [1 \; 1.4142]$.

Thus, the nonlinear feedback control law "u" is finally formulated as

$$u = V_{sc} = -\frac{\left(\frac{\omega_s P_M}{2H} - \frac{D(\omega(t) - \omega_s)}{2H}\right)}{-\frac{\omega_s E_q' V_\infty \sin \delta}{2HG_1}} + \frac{1}{-\frac{\omega_s E_q' V_\infty \sin \delta}{2HG_1}}[-(\delta(t) - \delta_0) - 1.4141(\omega(t) - \omega_s)] \tag{20}$$

2.3 Application of Noncooperative Game Theory in Nonlinear Controller Design

The desired optimal control problem for the power system under study can be treated under the noncooperative dynamic game theory. Following the theory of noncooperative game [10], the power system problem stated in (16) has been structured for time-varying case as

$$\dot{Z}(t) = f(t, Z(t), \gamma(t)); \quad Z(0) = Z_0 \tag{21}$$

and the linear quadratic type performance index is formulated as

$$J(t, \gamma(t)) = \frac{1}{2}Z^T(t_f)Q_f Z(t_f) + \frac{1}{2}\int_{t_0}^{t_f} \left(Z^T(t)QZ(t) + \gamma^T(t)R\gamma(t)\right) \tag{22}$$

Now from (21) to (22) an equation, known as *Isaacs Equation*, can be estimated as

$$-\frac{\partial V(t, Z(t))}{\partial t} = \min_{u_1 \in s_1}\left[\frac{\partial V(t, Z(t))}{\partial Z(t)} f(t, Z(t), \gamma^*(t, Z(t))) + g(t, Z(t), \gamma^*(t, Z(t)))\right] \tag{23}$$

The solution to this Eq. (23) is known as *Nash Equilibrium* point which can be derived as

$$\gamma(t) = -B^T(t)P(t)Z(t) \tag{24}$$

The above solution for time-invariant system becomes

$$\gamma(t) = -B^T P Z(t) \tag{25}$$

Finally, the desired control input "u" has been determined as [11]

$$u = V_{sc} = -\frac{\left(\frac{\omega_s P_M}{2H} - \frac{D(\omega(t) - \omega_s)}{2H}\right)}{-\frac{\omega_s E_q' V_\infty \sin\delta}{2HG_1}} + \frac{1}{-\frac{\omega_s E_q' V_\infty \sin\delta}{2HG_1}}[-(\delta(t) - \delta_0) - 1.7325(\omega(t) - \omega_s)] \tag{26}$$

This equation will be used later in Sect. 4 to investigate the performance analysis.

3 Stochastic Approach for the Design of Nonlinear Controller

3.1 LMI-Based H_∞ Excitation Controller Design

The standard mixed-sensitivity based feedback control associated with the SMIB test system can be demonstrated through Fig. 3. The closed-loop transfer function from disturbance input "d" to the controlled output "z" is derived as

$$T_{zd} = \begin{bmatrix} W_1(s)S(s) \\ W_2(s)K(s)S(s) \end{bmatrix} = C_c(sI - A_c)^{-1}B_c + D_c \tag{27}$$

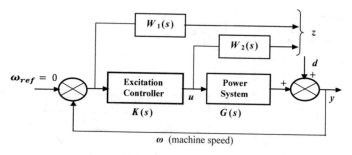

Fig. 3 Standard mixed-sensitivity based H_∞ control

where A_c, B_c, C_c, and D_c are matrixes of the closed-loop system; they have usual significances. The aim of the mixed-sensitivity based H_∞ control theory is to realize a controller $K(s)$ which is not only internally stable but also minimizes the transfer function between "d" to "z" which is represented by [12]

$$\|T_{zd}\|_\infty < \gamma \tag{28}$$

The problem under study thus referred to as the LMI-based H_∞ controller design is subject to the pole placement constraints with $\frac{\theta}{2} = 60°$. The said multi-objective H_∞ controller can be synthesized through the LMI Toolbox in MATLAB [13].

Finally, the constituted LMI-based H_∞ excitation controller is given by [8]

$$K(s) = \frac{0.0128s^2 + 6.6123s + 8.1334}{s^2 + 79.5909s + 11.6492}$$

3.2 Design of Conventional STATCOM Applying PSO- and GA-Based Techniques

In Fig. 4, the portrait of a PI-based traditional voltage controller has been presented [14]. It is worthwhile to mention that machine speed (ω) pertains the behavior of all the swing modes of the system. Figure 4 can be explored through the following set of differential equations:

$$\Delta \dot{X}_{s_2} = -\frac{1}{T_m} \Delta X_{s_2} + \frac{K_\omega}{T_m} \Delta \omega - \frac{1}{T_m} \Delta V_{meas} \tag{29}$$

$$\Delta \dot{X}_{s_3} = \left(-\frac{K_P}{T_m} + K_I\right) \Delta X_{s_2} + \frac{K_P K_\omega}{T_m} \Delta \omega - \frac{K_P}{T_m} \Delta V_{meas} \tag{30}$$

$$\Delta \dot{V}_{sc} = -\frac{1}{T_2} \Delta V_{sc} + \frac{1}{T_2} \Delta X_{s_3} + \frac{T_1}{T_2}\left(-\frac{K_P}{T_m} + K_I\right) \Delta X_{s_2} + \frac{T_1 K_P K_\omega}{T_2 T_m} \Delta \omega - \frac{T_1 K_P}{T_2 T_m} \Delta V_{meas} \tag{31}$$

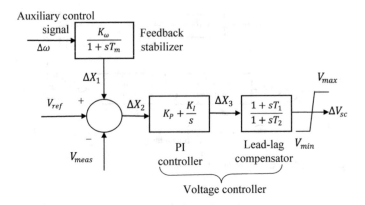

Fig. 4 Voltage controller block diagram for STATCOM

Furthermore, the power system given by (10)–(11) can be linearized by

$$\Delta\dot{\delta} = \Delta\omega \tag{32}$$

$$\Delta\dot{\omega} = -\frac{K_1\omega_s}{2H}\Delta\delta - \frac{D\omega_s}{2H}\Delta\omega - \frac{K_{V_{sc}}\omega_s}{2H}\Delta V_{sc} \tag{33}$$

where $K_1 = \frac{\partial P_e}{\partial \delta}$ and $K_{V_{sc}} = \frac{\partial P_e}{\partial V_{sc}}$. Therefore, (29)–(33) jointly accounted for the combined linear model of the test system with conventional controller. Four unspecified elements ($K_P, K_I, T_1,$ and T_2) of the traditional controllers are tuned up applying *Particle Swarm Optimization* (*PSO*) as well as by the *Genetic Algorithm* (*GA*) separately.

Particle Swarm Optimization (PSO) based Method

PSO is a metaheuristics approach first introduced in [15]. PSO-based optimization problem has been solved through minimization of a user-defined cost functional, named as Optimum Damping Index (*ODI*), which is presented by

$$ODI : \chi_i = (1 - \eta_i) \tag{34}$$

In (34), η_i indicates the damping ratio for ith oscillatory mode which is most critical. The η_i and hence "χ_i" can be evaluated through the iterative computation of the system matrix and eigenvalue of the conjugate linear model of the power system in association with the STATCOM controller.

Table 1 PSO-based results for linear STATCOM

Controller parameters	Limiting range (Min, Max)	PSO-based results	ODI (χ_i)
K_I	0.1, 5.0	2.0	0.4744
K_P	1.0, 10.0	3.33	
T_1	0.10, 1.50	1.50	
T_2	0.01, 0.25	0.21	

In order to execute the optimization problem in MATLAB, "*PSO toolbox*" has been utilized. In this problem "particle" has been represented by the following vector equation:

$$\text{Particle} = \begin{bmatrix} K_P & K_I & T_1 & T_2 \end{bmatrix} \tag{35}$$

The elements of (35) are composed of four tuning parameters of the STATCOM controller and the limiting values of these elements and PSO-based results are depicted in Table 1.

Optimization via Genetic Algorithm (GA)

The optimization process for the system described in Sect. 3.2 can also be done using GA [16]. The structure of the genome can be presented by the array of controller parameters:

$$\text{Genome} = \begin{bmatrix} K_P & K_I & T_1 & T_2 \end{bmatrix} \tag{36}$$

In this work, GA-based optimization problem has been solved in MATLAB utilizing "Genetic algorithm and direct search" Toolbox [17]. The optimal parameter set of the controller is evolved with the termination of the algorithm either at the iteration limit or at the desired fitness level of the objective function. The best individuals (controller parameters) and the minimum value of the fitness function "χ_i" are enlisted in Table 2.

Table 2 STATCOM parameters through GA

Parameters of the controller	Limiting range (Min, Max)	Results by GA	ODI (χ_i)
K_I	0.1, 5.0	0.3132	0.4229
K_P	1.0, 10.0	1.1263	
T_1	0.10, 1.50	0.2637	
T_2	0.01, 0.25	0.2469	

4 Performance Analysis of Nonlinear STATCOM Controller

To analyze the performance of the STATCOM nonlinear controller, simulation time is considered as 7 s. The occurrence of fault is at 1 s and the fault clearance time is assumed 1.2 s. The dynamic variation of the rotor angle and speed of the SMIB system with zero dynamic based and H_∞-based STATCOM controller have been compared and shown in Figs. 5 and 6, respectively.

The performance analysis of nonlinear exact linearization based STATCOM controller with PSO-based linear controller are displayed in Figs. 7 and 8. The function of Game Theory based controller with respect to the traditional controller generated by GA has also been illustrated in Figs. 9 and 10, respectively. Both the rotor

Fig. 5 Rotor angle with and without control

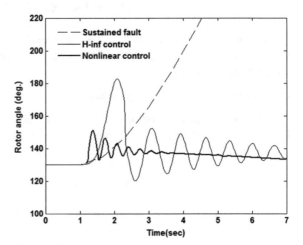

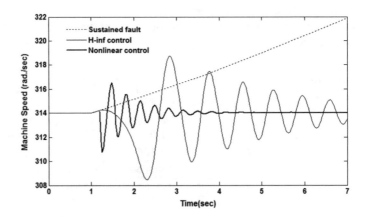

Fig. 6 Generator speed with and without control

Fig. 7 Response of rotor
angle variation

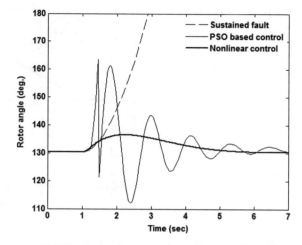

Fig. 8 Response of
generator speed variation

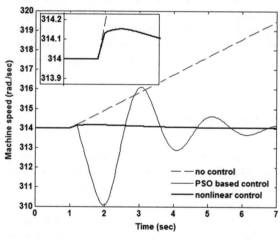

Fig. 9 Rotor angle behavior

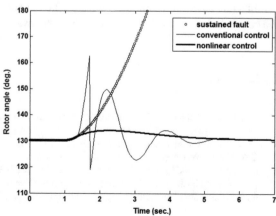

Fig. 10 Machine speed
behavior

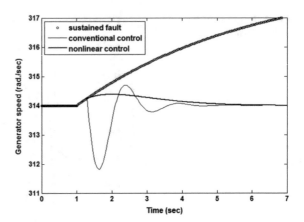

angle and machine speed response have been studied for typical fault scenario. It has been noticed that both the responses, rotor angle and the machine speed increase exponentially with sustained fault. However, it has been observed that both responses settle very quickly to the equilibrium position with the application of nonlinear-based STATCOM controller with respect to the conventional linear controller.

5 Conclusion

The proposed research work attempts to design nonlinear FACTS controller applying various methods to alleviate transient stability problems in power systems. The performances of the nonlinear controllers have been investigated in comparison to the conventional linear controller. The superiority and robustness of the nonlinear control over linear control have been shown through the application of typical contingency in the one-machine infinite-bus power system. This research work clearly exhibits that the nonlinear controllers are efficient and superior in greater extent than the conventional controller in mitigating power system stability problem.

References

1. A. Khodabakhshian, M.J. Morshed, M. Parastegari, Coordinated design of STATCOM and excitation system controllers for multi-machine power systems using zero dynamics method. Int. J. Electr. Power Energy Syst. (49), 269–279 (2013)
2. C.A. King, J.W. Chapman, M.D. Iliac, Feedback linearizing excitation control on a full-scale power system model. IEEE Tans. Power Syst. **9**(2), 1102–1109 (1994)
3. H.C. Tsai, J.H. Liu, CC. Chu, Integrations of neural networks and transient energy functions for designing supplementary damping control of UPFC. IEEE Trans. Ind. Appl. (2019)

4. C. Josz, D.K. Molzahn, M. Tacchi, S. Sojoudi, Transient stability analysis of power systems via occupation measures, in *IEEE Power & Energy Society Innovative Smart Grid Technologies Conference (ISGT)* (Washington, DC, USA, 2019), pp. 1–5

5. A. Halder, D. Mondal, N. Pal, Nonlinear optimal STATCOM controller for power system based on hamiltonian formalism, in *IEEE International Conference on Power Electronics, Drives and Energy Systems (PEDES)* (IIT Madras, India, 2018)

6. A. Halder, D. Mondal, M.K. Bera, Design of sliding mode excitation controller to improve transient stability of a power system, in *Association for the Advancement of Modeling and Simulation Techniques in Enterprises (MS-17)* (Kolkata, India, 2017)

7. Q. Lu, Y. Sun, S. Mei, *Nonlinear Control Systems and Power System Dynamics* (Kluwer Academic Publishers, Norwel, USA, 2001)

8. A. Halder, D. Mondal, *Zero Dynamics Design of Excitation Controller with Comparison to H∞ based design in Power System Stability Improvement* (Michael Faraday IET International Summit, Kolkata, India, 2015)

9. L. Gu, J. Wang, Nonlinear coordinated control design of excitation and STATCOM of power systems. Electr. Power Syst. Res. **77**, 788–796 (2007)

10. T. Basar, G.J. Olsder, *Dynamic Non cooperative Game Theory, Mathematics in Science and Engineering*, vol. 160 (Academic Press, 1982)

11. A. Halder, D. Mondal, Nonlinear optimal STATCOM controller based on game theory to improve transient stability, in *2nd International Conference on Control, Instrumentation, Energy and Communication* (Kolkata, India, 2016)

12. P. Gahinet, P. Apkarian, A linear matrix inequality approach to H_∞ control. Int. J. Robust Nonlinear Control. **4**(4), 421–448 (1994)

13. C. Scherer, P. Gahinet, M. Chilali, Multiobjective output-feedback control via LMI optimization. IEEE Trans. Autom. Control. **42**(7), 896–911 (1997)

14. K.R. Padiyar, *Analysis of Subsynchronous Resonance in Power Systems* (Kluwer Academic Publishers, Norwel, USA, 1999)

15. J. Kennedy, R. Eberhart, Particle swarm optimization, in *Proceedings of the IEEE International Conference on Neural Networks* (Perth, Australia, 1995), pp. 1942–1948

16. D. Goldberg, *Genetic Algorithms in Search, Optimization and Machine Learning*. Addison-Wesley, (1989)

17. MATLAB, Ver. 7.0, Genetic Algorithm Direct Search Toolbox, Version 1.0.1 (R14) (2004)

Towards Sustainable Indian Smart Cities: An AHP Study

Ankita Das, Abhijit Das and Nipu Modak

Abstract In this study, 12 parameters dictating the sustainability of Indian smart cities, identified using the literature review and field survey have been ranked using Analytical Hierarchical Process (AHP) nested with three sustainability parameters as criteria. The analysis shows that resource circulation, recycling rate, water management, e-waste management, health and safety, and job opportunity are the most important parameters. The sensitivity analysis reveals that resource circulation, recycling rate, health and safety, and job opportunity are the most sensitive parameters at different levels of sensitivity. The results will be helpful to the policy-makers and the researchers in developing sustainable smart cities in India.

Keywords Smart city · Sustainability · Analytical hierarchical process

1 Introduction

The phrase 'Smart city' has spread all over the world in the last decade which is impacting urban strategies in all sorts of towns [1, 2]. Smart city combines a variety of resources, technologies and administrative activities for the comfort of the citizens towards the goal of sustainable development [3]. Smart city is a fuzzy concept that has no specific definition [4]. There are several definitions of smart city which has been cited in the contemporary literature [4]. According to Giffinger [5] 'A Smart City is a city well performing built on the 'smart' combination of endowments and activities of self-decisive, independent and aware citizens'. The concept of smart city encompasses all the aspects needed for the citizens including basic needs, comfort and security (digital and reality) [6]. The execution process may vary based on

A. Das (✉)
Heritage Institute of Technology, Kolkata 700107, India
e-mail: ankitadas.ju@gmail.com

A. Das
RCC Institute of Information Technology, Kolkata 700015, India

N. Modak
Jadavpur University, Kolkata 700032, India

© Springer Nature Singapore Pte Ltd. 2020
S. Bhattacharyya et al. (eds.), *Intelligence Enabled Research*,
Advances in Intelligent Systems and Computing 1109,
https://doi.org/10.1007/978-981-15-2021-1_11

geographical location, attitude of common people, sociopolitical agenda, strategic decision-making and most importantly optimal resource utilization [5, 6].

Smart city became a very popular phrase in India after the 'Smart City Mission' was initiated by the government of India on 25th June 2015 [6, 7]. The mission is orchestrated by the Ministry of Housing and Urban Affairs (MoHUA), Government of India [7]. The mission focuses on developing 100 smart cities in India by 2021 [8]. The said mission has been objectivized to create models that can be further customized, reused and replicated to create and develop other smart cities in various regions and parts of the country [8]. The key features of a Smart City include adequate water supply, availability of electricity, proper sanitation, sustainable waste management, efficient urban mobility, reasonable housing, healthy ICT implementation and connectivity, health and safety of citizens, green economy, cyber security, education and equity for citizens [8, 9].

India is gearing up for the development of smart cities which is laudable, but the question that comes up at this moment is whether the cities will be only smart cities. According to UNECE–ITU [10], 'A smart sustainable city is an innovative city that uses information and communication technologies (ICTs) and other means to improve quality of life, efficiency of urban operation and services, and competitiveness, while ensuring that it meets the needs of present and future generations with respect to economic, social, cultural and environmental aspects'. Hence, it is also imperative to keep sustainability in mind. In this study, total of 12 parameters of sustainable smart city development have been considered specific to Indian conditions. Analytical Hierarchical Process (AHP), a multi-criteria decision-making (MCDM) technique was used to prioritize the chosen parameters with respect to three pillars of sustainability, i.e. environmental, economic and social. Additionally, a sensitivity analysis was carried out to test the sensitive parameters.

2   Study Methodology

The study was carried out in two phases. In phase one, a detailed literature review was carried out with keywords such as 'smart city', 'sustainable smart city', 'smart city in India' and 'smart city indicators'. This process was carried out to churn out the relevant information from the reported studies. Thereafter, a field study was carried out in three metro cities in India with semi-structured questionnaire to understand the view of common people as well as academic practitioners towards smart city and sustainability. Both the above-mentioned exercises were used to shortlist the parameters which are important towards the journey of becoming sustainable smart cities. In the second phase of the study, a hierarchical model (Fig. 1) was developed following the principles of AHP developed by Saaty [11]. AHP was chosen as it offers a robust interface with less computational complexity over other MCDM techniques such as TOPSIS and SAW. The goal was to find the most important parameters towards the journey of sustainability for the smart cities nested with the three pillars of sustainability as criteria and 12 chosen parameters as the alternatives. The AHP

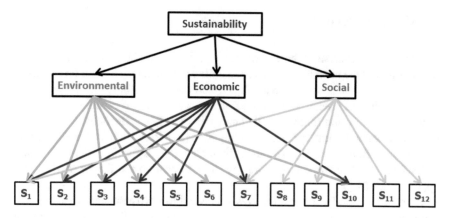

Fig. 1 AHP model for sustainable smart city

analysis has been carried out using the 'Super Decision' software (version 2.10.0, released 14 January 2019) [12]. The operational algorithm of AHP coined by Saaty [11] has been elaborated below.

Step 1: Problem definition and identification of the goal, criteria and alternatives to structure the hierarchy model.

Step 2: Construction of n × n comparison matrices for each of the subordinate levels of goal with one matrix for each node in the level directly above the concerned level for pairwise comparison. The pairwise comparisons are carried out to identify which overpowers the other nodes.

Step 3: A total n (n −1)/2 number of judgmental ratings are required to develop each set of matrices in step 2. Reciprocals get assigned inevitably in each of the pairwise comparisons.

Step 4: The eigenvectors are weighed by the weights of the criteria using hierarchical synthesis. For each element in the next subordinate level, the summation is done over all weighted eigenvector entries.

Step 5: After carrying out all pairwise comparisons, the consistency of the matrix is determined by using the eigenvalue, λ_{max}, to calculate the consistency index, CI as follows:

$$CI = (\lambda_{max} - n)/(n - 1) \qquad (1)$$

where n is the matrix size. Overall consistency is validated by evaluating the consistency ratio (CR) of CI with the appropriate value. If the value of CR does not exceed 0.10, then it is acceptable, otherwise the matrix is inconsistent.

$$CR = CI/\text{Random Consistency} \qquad (2)$$

Step 6: Steps 2–5 are performed for all levels in the hierarchy.

3 Model Description

The AHP model consists of three levels of hierarchy. The highest level is the goal level which is to find the most important sustainability parameters. The next level consists of three criteria nodes which are the three pillars of sustainability, i.e. Environmental, Economic and Social. The last level consists of alternatives, i.e. the 12 chosen parameters—New and Innovative Technologies (S_1), Water Resource Management (S_2), Wastewater Treatment (S_3), Recycling Rate (S_4), Repair and Refurbishing Rate (S_5), E-waste Recycling (S_6), Resource Circulation (S_7), Cyber Security (S_8), Health and Safety (S_9), Climate Change Strategy (S_{10}), Education and Equity ($S_{11)}$ and Job opportunities (S_{12}).

In the above model, the alternatives connected to environmental node are New and Innovative Technologies, Water Resource Management, Wastewater Treatment, Recycling rate, Repair and Refurbishing Rate, E-waste Recycling, Resource circulation and Climate Change Strategy. The economic criteria node is connected to New and Innovative Technologies, Water Resource Management, Wastewater Treatment, Recycling rate, Repair and Refurbishing Rate, Resource circulation and Climate Change Strategy. The social criteria node is connected to New and Innovative Technologies, Resource Circulation, Cyber Security, Health and Safety, Education and Equity and Job opportunities.

4 Results and Discussion

The AHP analysis gives us results in two folds, i.e. it gives the prioritized alternatives individually with respect to individual criteria as well as rankings with respect to the goal of the analysis. With respect to the environmental sustainability, Water Resource Circulation, Wastewater treatment, E-waste recycling and Resource Circulation are the most important parameters. Considering the economic sustainability, Resource Circulation (0.358), Recycling Rate (0.229), Repair and Refurbishing Rate (0.141) and New and Innovative Technologies (0.104) are the most important alternatives. Analysis with respect to social sustainability reveals that Health and Safety (0.348), Job opportunity (0.240) and Education and Equity (0.166) are the most important parameters. The AHP analysis is generally carried out based on ratings given by experts. The expert opinions can be biased which can compromise the decision-making process. Here three cases were chosen to eliminate those possibilities. Since the criteria of the AHP models are the three pillars of sustainability, three cases can arise—(a) Environmental Sustainability gets the highest priority (environmentalist's perspective), (b) Economic Sustainability gets the highest priority (economist's and industrialist's perspective) and (c) Social Sustainability gets the highest priority (socialist's perspective) and they are discussed in detail.

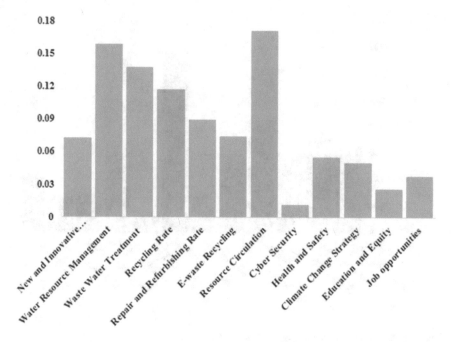

Fig. 2 Ranked alternatives with high environmental priority

4.1 Case Study 1: High Environmental Priority

Figure 2 shows the ranking of the alternatives with respect to environmental sustainability. In this case, resource circulation, water resource management and wastewater treatment are the most important parameters. Whereas the least important parameters are job opportunity, education and equality and cyber security. It may be noted that recycling rate, repair and refurbish rate and e-waste recycling occupy significant areas in the plot which implies that these should be given additional priority in action plans and policy frameworks.

4.2 Case Study 2: High Economic Priority

Figure 3 shows the ranking of the alternatives with respect to economic sustainability. In this case, resource circulation; recycling rate and water resource management are the most important parameters. Whereas the least important parameters cyber security, e-waste recycling and climate change strategy. It may be noted that recycling rate, repair and refurbish rate and wastewater treatment occupy significant areas in the plot which implies that these should be given additional priority in action plans and policy frameworks.

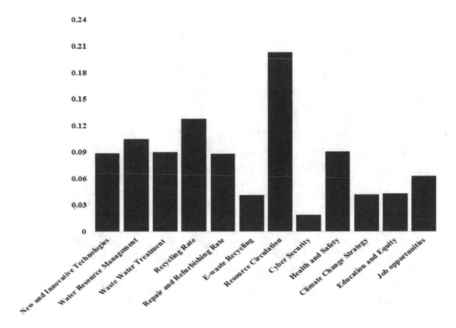

Fig. 3 Ranked alternatives with high economic priority

4.3   Case Study 3: High Social Priority

Figure 4 shows the ranking of the alternatives with respect to the social sustainability. In this case, health and safety, job opportunity and resource circulation are the most important parameters. Whereas the least important parameters e-waste recycling, cyber security and climate change strategy. It may be noted that new and innovative technologies, recycling rate water resource management and wastewater treatment occupy significant areas in the plot which implies that these should be given additional priority in action plans and strategic frameworks.

4.4   Sensitivity Analysis

Sensitivity analysis was carried out using the said software. In this case, the priorities of all three criteria were kept equal such that there is no effect of biasness on the sensitivity results. Figure 5 shows the sensitivity plot of the alternatives with respect to the environmental criteria. It is clear from the plot that job opportunity, resource circulation and recycling rate are affected the most when medium sensitivity is revoked.

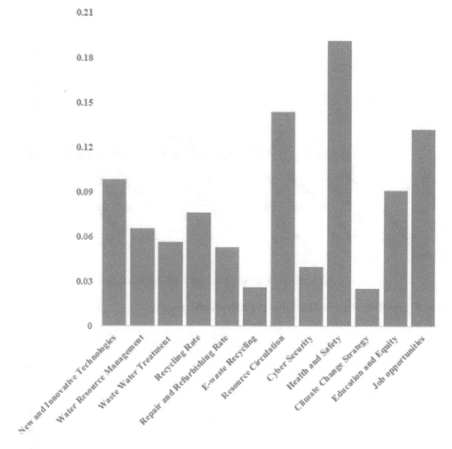

Fig. 4 Ranked alternatives with high social priority

Figure 6 shows the sensitivity plot of the alternatives with respect to the environmental criteria. It is clear from the plot that job opportunity, resource circulation, and health and safety is affected the most when medium sensitivity is revoked. Whereas in the highest sensitivity, recycling rate and resource circulation are the most affected parameters.

Figure 7 shows the sensitivity plot of the alternatives with respect to the environmental criteria. It is clear from the plot that job opportunity and resource circulation are affected the most when medium sensitivity is revoked. Whereas in the highest sensitivity, health and safety; education and equity and job opportunity are the most affected parameters. Even at no sensitivity, resource circulation is affected.

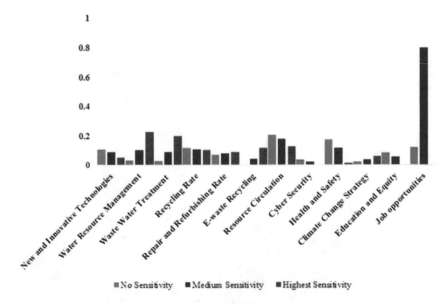

Fig. 5 Sensitivity analysis with respect to environmental criteria

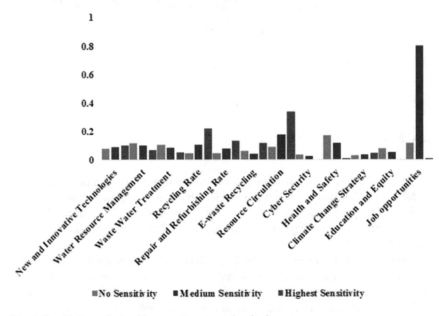

Fig. 6 Sensitivity analysis with respect to economic criteria

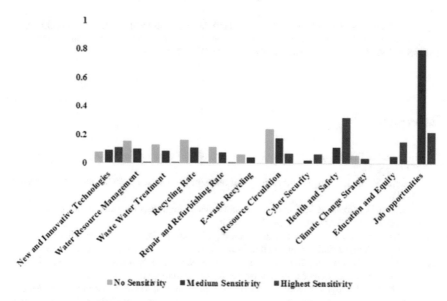

Fig. 7 Sensitivity analysis with respect to social criteria

5 Conclusion

The buzz of smart city has gone viral among the urban dwellers in the last 4 years and it is going to stay there for a long time. From a very basic level of understanding, smart city is a technology-enabled environment by the people living in it for the people who live. Despite all this, the arrangements must be sustainable in the long run. In this study, the focus is on the sustainability of Indian smart cities. The factors governing the sustainability of Indian smart cities were derived from the literature study and field survey which were used to develop an analytic hierarchical model in the veracity of three sustainability pillars. Three bias cases with respect to the sustainability pillars were presented to remove biasness towards expert opinions. The AHP analysis revealed that resource circulation, recycling rate, repair and refurbish rate, water resource management, e-waste management, health and safety, and job opportunity are the most important factors overall, which should be given utmost importance for developing sustainable smart cities in India. The sensitivity analysis was carried out at levels of no, medium and high sensitivity of environmental, economic and social parameters. It reveals that resource circulation, recycling rate, health and safety, and job opportunity are the most sensitive parameters at different levels of sensitivity. Job opportunity and recycling rate were affected the most in all three cases. This study might be constricted within the biasness of identification of parameters. However, more studies of such calibre may prove to be beneficial as this one can serve as a prima facie to develop models of higher complexity and dimension such as a decision support system.

Acknowledgements The author would like to thank Mr. Biswajit Debnath, Commonwealth Split-site Scholar, Aston University, for his constant support, encouragement and help during the AHP analysis and guidance towards learning the SuperDecision Software.

References

1. R.P. Dameri, C. Rosenthal-Sabroux, Smart city and value creation, *Smart City* (Springer, Cham, 2014), pp. 1–12
2. A. Caragliu, C. Del Bo, P. Nijkamp, Smart cities in Europe. J. Urban Technol. **18**(2), 65–82 (2011)
3. V. Albino, U. Berardi, R.M. Dangelico, Smart cities–definitions, dimensions, and performance, in *Proceedings IFKAD* (2013), pp. 1723–1738
4. R.P. Dameri, Comparing smart and digital city: initiatives and strategies in Amsterdam and Genoa. Are they digital and/or smart?, *Smart City* (Springer, Cham, 2014), pp. 45–88
5. R. Giffinger, N. Pichler-Milanović, Smart cities: ranking of European medium-sized cities (Centre of Regional Science, Vienna University of Technology, 2007)
6. N.M. Kumar, S. Goel, P.K. Mallick, Smart cities in India: features, policies, current status, and challenges, in *2018 Technologies for Smart-City Energy Security and Power (ICSESP)* (2018, March). IEEE, pp. 1–4
7. smartcities.gov.in (2019), http://smartcities.gov.in/. Accessed 12 Sept 2019
8. Mohua.gov.in., Smart cities: Ministry of Housing and Urban Affairs, Government of India (2019), http://mohua.gov.in/cms/smart-cities.php. Accessed 12 Sept 2019
9. P. Bosch, S. Jongeneel, V. Rovers, H.M. Neumann, M. Airaksinen, A. Huovila, CITYkeys indicators for smart city projects and smart cities. CITYkeys report (2017)
10. UNECE (2015), http://www.unece.org/fileadmin/DAM/hlm/projects/SMART_CITIES/ECE_HBP_2015_4.pdf. Accessed 12 Sept 2019
11. T.L. Saaty, The analytic hierarchy process (McGraw-Hill. New York, 1980), p. 324
12. Superdecisions.com, https://www.superdecisions.com/. Accessed 13 Sept 2019

Temporal Sentiment Analysis of the Data from Social Media to Early Detection of Cyberbullicide Ideation of a Victim by Using Graph-Based Approach and Data Mining Tools

A. Chatterjee and A. Das

Abstract There are lots of works that can be found on sentiment analysis by using various data mining tools. In most of the cases, product evaluations, measurement of sentiment polarity, detection of illness, etc., have been done by this method of analysis. In this paper, we have taken an approach of graph-based analysis of sentiments from social media like Twitter to detect and prevent the cyberbullicide ideation.

Keywords Twitter · Cyberbullicide · Suicide · Sentiment · Data mining · Tweepy · Word cloud · Text blob · Temporal · Sentiment polarity

1 Introduction

In this study, we have taken Twitter—which is one of the widest used and most popular microblogging sites, which allows its users to communicate using short-length messages. Contrasting to the other available social media, here in Twitter the relationships among Twitter users are likely to be two—one is the followee and the other is the follower. A user turns out to be a follower when he adds another as a contact or friend while the other will be a followee. Finally, when a user publishes a tweet, then this spontaneously appears on his home page as well as the home pages (user timeline) of his followers [1]. Cyberbullicide is suicide by a person due to cyber harassment, rumors spread over online social media, and other forms of cyberbullying. It has become a global problem and younger generations are mostly victimized because they are much more active than the relatively older generations in cyberworld socially [2, 3]. But that does not decrease the probability of relatively older generations. So the victim can be of any age. Our objective is early detection

A. Chatterjee (✉)
Techno India Hooghly, West Bengal, India
e-mail: avikchatterjee@technoindiahooghly.org

A. Das
Department of IT, RCCIIT, West Bengal, India
e-mail: ayideep@yahoo.co.in

© Springer Nature Singapore Pte Ltd. 2020
S. Bhattacharyya et al. (eds.), *Intelligence Enabled Research*,
Advances in Intelligent Systems and Computing 1109,
https://doi.org/10.1007/978-981-15-2021-1_12

of the suicidal tendency of a person due to cyberharassment and finally to prevent a person from this fatal act of self-destruction. We have choose Twitter as it is easy to access by various data mining tools [4–9].

2 Related Work

Identifying which tweets defines an explicit event and clustering them properly is one of the main challenging tasks related to social media currently addressed in the NLP community. Existing approaches to extract events from tweets can be classified into two main classes, namely closed-domain and open-domain event identification systems. In the closed-domain, approaches are mainly focused on extracting a particular type of event, as for instance natural disasters. Works in the closed-domain situation are usually cast as supervised classification tasks that trust in keywords to extract event-related messages from Twitter. The open-domain situation is more challenging, as it is not restricted to an explicit type of event and usually trusts on unsupervised models [10]. Here in this paper, we have taken keywords related to cyberbullicide and k terms that precede and follow their mention in a tweet and a graph are prepared.

3 Our Approach

Keywords related to cyberbullicide are searched by various Python data mining tools like Tweepy and Textblob. The searched keyword and k terms that precede and follow their mention in a tweet is considered as nodes. Nodes in the graph are connected by a directed edge if they co-occur in the context of a related keyword. Let $G = (V, E)$ be a graph where V is the set of vertices and E is the set of edges. Here vertex V_i and V_j are the keywords related to twitter handler i and j, respectively, where $i \neq j$. The weight of the edge between vertex V_i, V_j is denoted by ω_{ij}. The value of ω_{ij} is measured by the average polarity of V_i and V_j within a specified time window (t_i, t_j). t_i, t_j are the creation time stamps of vertex V_i and vertex V_j, respectively. Let ε_{ij} be the edge between V_i and V_j. There are three types of polarity, positive, negative, and neutral indicated as 1, -1, and 0, respectively. We can calculate by our algorithm that the weight (ω_{ij}) of any edge ε_{ij} lies in the range of 1–0. If α is the threshold weight of any edge ε_{ij}, then the permissible range of α is $0 \leq \alpha \leq 1$. Let $In(V_i)$ be the set of vertices that points to V_i (i.e., predecessors), and $Out(V_i)$ be the set of vertices that V_i points to (i.e., successors). So, ε_{ij} is $Out(V_i)$ which points to V_j. Let r_{ij} be the risk factor for V_i which is determined by the ω_{ij} of $Out(V_i)$. So, $r_{ij} = f(\omega_{ij})$, where, $\omega_{ij} = (pol(v_i) + pol(v_j))/2$. In this equation $pol(V_i)$ and $pol(V_j)$ are the sentiment polarities of vertex V_i and V_j, respectively. In this paper, we have followed the Temporal Analysis of graph-based approach. The lesser the weight of the successor($out(V_i)$), the risk r_{ij} of cyberbullicide decreases.

In Fig. 1, V_h, V_i, and V_j are different keywords that belong to different twitter handles. In the graph, these are treated as nodes. The ε_{hi} and ε_{ij} are the edges between V_h, V_i, and V_i, V_j, respectively. Time window indicates the period (time frame) of data collection. In this way, the graph will grow longer with consideration of more keywords related to the same topic.

4 Algorithm

4.1 Algorithm for Cyberbullicide Detection

Let Polarity of $V_i = pol(v_i)$ and Polarity of $V_j = pol(v_j)$ within any time window (t_i, t_j), ω_{ij} is the weight of vertex V_i and V_j, r_{ij} is the risk factor associated with V_i, α is the threshold weight of any age ε_{ij}.

if $\omega_{ij} = (pol(v_i) + pol(v_j))/2$ **then**
| $r_{ij} = \sum_i^j f(\omega_{ij})$ for any edge ε_{ij}
if $\omega_{ij} \notin \alpha$ **then**
| r_{ij} is high and the twitter handle i need to take care
else
| r_{ij} is low and no care is needed

Figures 2, 3, 4, and 5 show the word cloud implementation of different keywords related to cyberbullying.

5 Result and Discussion

We can gather the data by using Tweepy and Textblob tools of Python from Twitter. We can also use n-gram and word cloud tools to find the keywords. Then the sentiment polarity of the keywords is calculated. Based on this we can prepare the graph and assign weights to the edges of the graph. After this, we can apply our algorithm to detect the cyberbullicide tendency for a particular twitter handle. If "High Risk" is detected for any twitter handle, then proper care can be taken for such a user.

The data in Table 1 is collected from different Twitter accounts with the help of Python data analysis tools. The keywords chosen in Table 1 are just a few examples.

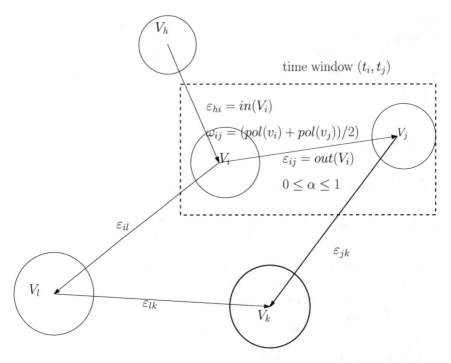

Fig. 1 Example graph

Fig. 2 Word cloud implementation 1

Fig. 3 Word cloud implementation 2

Fig. 4 Word cloud implementation 3

Fig. 5 Word cloud implementation 4

Table 1 Polarity of a few common terms used by cyberbullies extrated from Twitter by using Python tools

Keyword	Positive%	Negative%	Neutral%
Dumb	21	75	3
Kill	20	64	15
Maim	20	12	66
Destroy	20	45	34
Loser	20	12	67
Fatso	31	14	53
Nerd	26	25	48
Ugly	8	82	8
Slut	28	21	50
Freak	39	17	42
Jerk	30	32	36

5.1 n-gram Approach

| sentence | = | 'You | are | a | loser, | I | will | ruin | you' |

('You',	'are',	'a',	'loser,',	'I',	'will')
('are',	'a',	'loser,',	'I',	'will',	'ruin')
('a',	'loser,',	'I',	'will',	'ruin',	'you')

| sentence | = | 'Youre | too | ugly | to | be | making | such | comments' |

('Youre',	'too',	'ugly',	'to',	'be',	'making')
('too',	'ugly',	'to',	'be',	'making',	'such')
('ugly',	'to',	'be',	'making',	'such',	'comments')

This section describes the n-gram approach to the sentences collected from Twitter for the purpose of polarity analysis.

In this way, we can collect different keywords related to cyberbullicide from different twitter handle and analyze them by using different Python data analysis tools.

6 Conclusion and Future Scope

Here we have concentrated on cyberbullicide ideation by temporal graph-based approach. This approach can also be applied to many different topics. Later, we can elaborate the work on suicidal ideation and other psychological problems like Stress,

Neurosis, Hypochondriasis, Somatization Disorder, Factitious Disorder, Schizophrenia, Paranoia, Depression, Mania, Manic Depression, Dementia, and Narcissism. In this work, we have not specified any particular age group or gender or profession or such kind of things. Later on, we can work to different gender or age group people to detect the symptoms more accurately. Besides this, we can apply the concept of graph-based approach on different product review systems and feedback analysis systems also.

References

1. Y. Vassiliou, D.P. Karidi, Y. Stavrakas, A personalized tweet recommendation approach based on concept graphs
2. S. Tonelli, N. Le-Thanh, A. Edouard, E. Cabrio, Graph-based event extraction from twitter
3. A.F. Rozie, A. Arisal, H. Nurrahmi, R. Wijayanti, Twitter data transformation for network visualization based context analysis
4. A. Khan, K. Khan, B.B. Baharudin, F. e Malik, Automatic extraction of features and opinion oriented sentences from customer review. World Acad. Sci. Eng. Technol. **62** (2010)
5. Y.-T. Liang, L.-W. Ku, H. Chen, Opinion extraction, summarization and tracking in news and blog corpora. AAAI-CAAW **06**
6. R. Narayanan, K. Zhang, Voice of the customers: mining online customer reviews for product (2010)
7. D.D. Wu, N. Li, Using text mining and sentiment analysis for online forums hotspot detection and forecast. Decis. Support. Syst. **48**, 354–368 (2010)
8. S. Skiena, N. Godbole, M. Srinivasaiah, Largescale sentiment analysis for news and blogs, in *Proceedings of ICWSM*, Boulder, Colorado, USA, 2007
9. R. Law, Q. Ye, Z. Zhang, Sentiment classification of online reviews to travel destinations by supervised machine learning approaches. Expert Syst. Appl. **36** (2009)
10. A. Pak, P. Paroubek, Twitter as a corpus for sentiment analysis and opinion mining, in *Proceedings of LREC*, vol. 10 (2010), pp. 132–136

Intelligent Color Image Watermarking Using Phase Congruency

Subhrajit Sinha Roy, Abhishek Basu and Avik Chattopadhyay

Abstract In this paper, an invisible color image watermarking technique has been developed as a copyright protection tool. This proposed scheme embeds a color watermark into a color cover image. Maximum watermark data is implanted into the most informative region and the amount of hidden data gradually decreases with lesser informative regions. The fragmentation of regions according to their business is performed through a phase congruency based feature map, followed by the generation of an intelligent hiding capacity map. The experimental results can show that this adaptive data embedding process is able to deal with the tradeoffs among imperceptibility, robustness, and data hiding capacity.

Keywords Color image · Digital watermarking · Payload · HVS · Imperceptible · LSB · Phase congruency

1 Introduction

Digital watermarking is a conventional art of multimedia copyright protection that embeds copyright information, known as a watermark, into cover objects either in spatial or in frequency domain. Usually, most of the watermarking techniques have been implemented by using images as the test cover objects as well as the watermark. Spatial domain offers higher imperceptibility and less system complexity, whereas frequency domain can provide high robustness. Least significant bit (LSB) replacement, LSB matching, correlation-based methods, and patchwork techniques are some of the habitual spatial domain watermarking practices. On the other hand, some basic signal transformation techniques, like discrete Fourier transform (DFT), discrete cosine transform (DCT), and discrete wavelet transform (DWT) are generally used as the fundamental tools for frequency-domain techniques, and thus,

S. Sinha Roy (✉) · A. Basu
RCC Institute of Information Technology, Kolkata, India
e-mail: subhrajitkcs@gmail.com

S. Sinha Roy · A. Chattopadhyay
University of Calcutta, Kolkata, India

© Springer Nature Singapore Pte Ltd. 2020
S. Bhattacharyya et al. (eds.), *Intelligence Enabled Research*,
Advances in Intelligent Systems and Computing 1109,
https://doi.org/10.1007/978-981-15-2021-1_13

these are often said as transform domain techniques. During the last few decades, a good number of image watermarking schemes have been generated in both of the domains to advance the performance. Introduction of human visual system (HVS) based techniques, singular value decomposition concept, etc., result in enhancement of system proficiency [1, 2]. Yet, it is hard to improve robustness, imperceptibility, and hiding capacity or payload simultaneously, as these three essential properties of any good watermarking system conflict with each other. This still makes the field of watermarking an open area for research.

Till now, most of the image watermarking methods are focused on embedding binary or grayscale watermark into grayscale or color images; but, in this paper, color images are considered for both, the cover object and the watermark. In other word, this proposed scheme embeds a color watermark into color images. Phase congruency has been involved here to generate cover image feature map, which is nearly akin to the feature identification process through HVS. Intelligent techniques are introduced to generate hiding capacity map to perform an adaptive data embedding process so that maximum data can be embedded into the most informative areas. In this way, data transparency of the watermark can be improved with higher payload.

The proposed methodology for watermark embedding and extracting has been demonstrated in Sect. 2. In Sect. 3, the experimental results have been provided, and proficiency of the proposed scheme has been compared to some existing image watermarking frameworks. Finally, this paper has been wrapped up in Sect. 4.

2   Proposed Methodology

In this section, an image watermarking method has been proposed to embed a color watermark into a cover image, which is also color in nature. Naturally, the watermarking process has two sections—watermark embedding and watermark extracting.

2.1   Watermark Embedding

Watermark embedding is performed through the three following steps using this proposed methodology.

Generation of phase congruency based feature map: In the first step, the color cover image is decomposed into three color planes (red, green, and blue) and the feature maps for each and every color plane are generated independently, using the concept of phase congruency. Phase congruency is introduced to utilize the flawed nature of HVS. In this paper, the phase congruency technique, proposed by Kovesi [3], is used to generate the feature map of each color plane of the cover image to distinguish image pixels according to their relative information quantity in that particular image plane.

Generation of hiding capacity map (HCM) using intelligent bit-depth estimation: The feature map of each color plane consists of data points having real values within the range of 0–1. Higher values in the feature map indicate more informative regions in the corresponding color plane. Now, HVS is less able to detect any slightest change in regions having a lot of information with respect to that in flat regions. To utilize this fact, an intelligent image clustering algorithm can be applied to the feature map to segment the cover image planes into a few distinct sets. Here, the fuzzy C-means (FCM) algorithm [4] has been followed to sector the cover image pixels of each color plane into four distinct regions. Each region consists of distinct mean value. Higher mean value indicates the regions with more informative pixels, where more amount of watermark information could be embedded imperceptibly, according to the nature of HVS. Thus, from feature map, a corresponding HCM is generated, where pixels with the highest value (brightest region) allow to embed maximum data in their corresponding cover image pixels, and for pixels with the lowest value (darkest region), embedding payload is the least. In this way, the adaptive bit depth for each cover image pixel is estimated and the maximum hiding capacity can be calculated. While generating watermark, it should be verified that the applied payload must be less than the maximum hiding capacity.

Watermark insertion through adaptive LSB replacement: This embedding process embeds the bits of the watermark pixels of red, green, and blue planes, respectively, into the pixels of the red, green, and blue planes of the cover image. An adaptive LSB replacement is followed to implant watermark bits into cover pixels. Based on the HCM pixel values, 1–4 numbers of LSBs of the cover image pixels are replaced by the corresponding watermark bits. The embedding process is performed in such a way that starting from the MSB, the higher bit planes' watermark information are embedded into higher bit planes (among the first four LSBs) of the cover image pixels so that, higher robustness could be achieved. This way, watermark embedding is performed for each color plane pair of the cover image and watermark. Finally, the actual watermarked color image is composed of these three watermarked color planes. The block diagram of this proposed watermark embedding scheme is given in Fig. 1.

2.2 Watermark Extracting

This proposed scheme is a non-blind technique, as it requires the HCM of the original cover image to extract the watermark from a received watermarked image. Based on the HCM pixel values of each color plane of the received image, first 4–1 number of LSBs are adaptively extracted and stored in a variable matrix such that the arrangement of these extracted bits are from higher bit plane to lower bit plane. Thus,

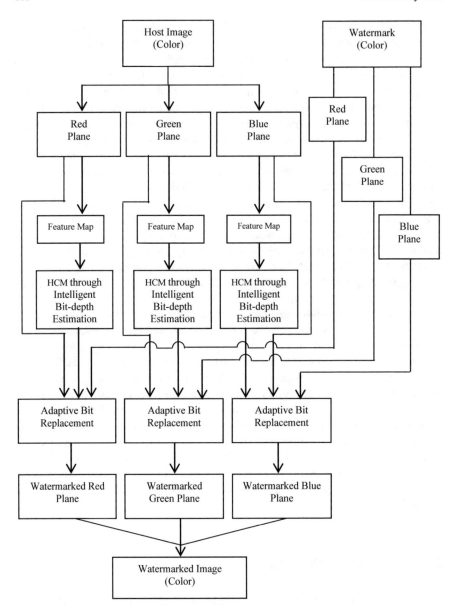

Fig. 1 Block diagram for watermark embedding system

the watermark bits can be found in order from MSBs to LSBs of all the watermark
pixels. Now, from the variable matrix, the watermark pixels of each color plane can
be reformed easily, and all three color planes together form the extracted watermark.
This extracted watermark is compared to the original one for testing the authenticity.
Figure 2 provides the block diagram of the proposed watermark extracting scheme.

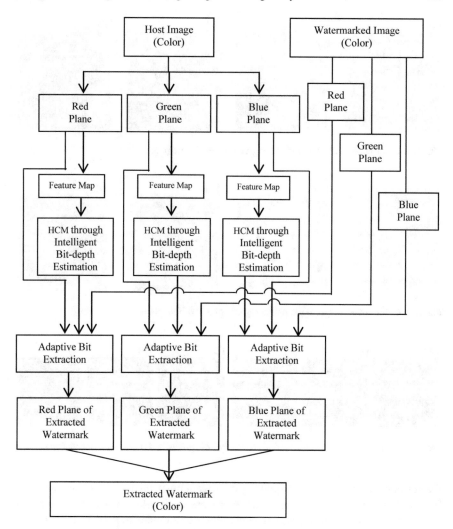

Fig. 2 Block diagram for watermark extracting system

3 Results and Discussion

The experiment on this proposed scheme has been carried out using a color watermark of size 96 × 96 and a set of color cover images (taken from an image database [5], conventionally used in image processing) of size 256 × 256. Figures 3 and 4 show the original watermark and one of the cover images (randomly chosen to set an example) along with their corresponding decomposed images for red plane, green plane, and blue plane, respectively. The feature maps for individual color planes of the cover image, generated through phase congruency, are shown in Fig. 5a–c, and

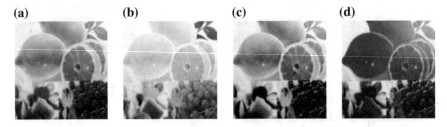

Fig. 3 **a** Color watermark; **b** red plane; **c** green plane; **d** blue plane

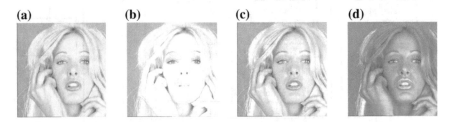

Fig. 4 **a** Color cover image; **b** red plane; **c** green plane; **d** blue plane

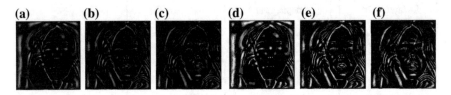

Fig. 5 **a–c** Phase congruency maps; **d–f** HCMs for red, green, and blue cover planes, respectively

the consequential HCMs after applying intelligent bit-depth estimation are shown in
Fig. 5d–f. As described in the embedding process, each color plane of the watermark
is embedded into the corresponding cover image color plane to produce the resultant
color plane for the watermarked image. The watermarked image with its decomposed
color planes is given in Fig. 6. There is hardly any visual distortion perceived in the

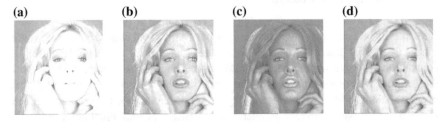

Fig. 6 **a–c** Watermarked red, green, and blue planes; **d** watermarked color image

images of Fig. 6 compared to the analogous images of Fig. 4. This reveals that the imperceptibility of the proposed scheme is high enough.

A set of image quality metrics [6] is employed here for quantitative analysis of system results in terms of imperceptibility and robustness, which are provided in Tables 1 and 2, respectively. From Table 1, it is found that the average peak signal-to-noise ratio (PSNR) is 44.7 dB. Moreover, the values of structural similarity index (SSIM), normalized cross-correlation (NCC), and universal image quality index (UIQI) affirm that the imperceptibility significantly is high, as the values of these metrics are found unity for two identical images.

Table 1 Results of system performance in terms of imperceptibility

Images	PSNR	MD	NC	SSIM	UIQI
4.1.01.tiff	44.47698	15	0.998053	0.988415	0.985128
4.1.02.tiff	45.60168	15	0.996074	0.991413	0.966239
4.1.03.tiff	46.25473	15	0.99987	0.990734	0.99999
4.1.04.tiff	43.4966	15	0.998893	0.985707	0.999726
4.1.05.tiff	44.55247	15	0.999672	0.988524	0.999986
4.1.06.tiff	44.84239	15	0.999519	0.992561	0.999974
4.1.07.tiff	43.7125	15	0.999835	0.982327	0.99996
4.1.08.tiff	43.24296	15	0.999577	0.982217	0.999961
4.2.01.tiff	44.26559	15	0.9995	0.984543	0.993049
4.2.02.tiff	43.83535	15	0.999684	0.985546	0.999987
4.2.03.tiff	48.66517	15	0.999739	0.998202	0.999998
4.2.04.tiff	43.43073	15	0.99969	0.983237	0.99994
4.2.05.tiff	45.21486	15	0.999676	0.992975	0.999968
4.2.06.tiff	45.187	15	0.999561	0.991907	0.999745
4.2.07.tiff	42.84179	15	0.999647	0.981272	0.99367
house.tiff	45.4816	15	0.999843	0.994953	0.999928

Table 2 Results of system performance in terms of robustness

Attacks	PSNR	NC	SSIM	UIQI
No. attack	∞	1	1	1
90° rotation	∞	1	1	1
15° rotation	11.19679	0.895289	0.102319	0.917067
Salt and pepper	25.03701	0.995945	0.766718	0.999079
LSB inversion (first LSB ↔ second LSB)	18.65367	0.982919	0.726222	0.989248
Negative	4.734683	0.518603	−0.40531	0.477455

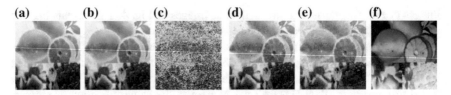

Fig. 7 Extracted watermark after applying attacks **a** no attack; **b** 90° rotation; **c** 15° rotation; **d** salt and pepper; **e** LSB inversion (first LSB ↔ second LSB); **f** negative

Table 3 Comparison of system proficiency with other existing frameworks

Attacks	PSNR (dB)	Payload (bpp)
Proposed scheme	44.69	1.27
Scheme 1 [7]	35	0.5
Scheme 2 [8]	50	0.02
Scheme 3 [9]	50	0.001
Scheme 4 [10]	52	0.04
Scheme 5 [11]	49	0.125

Maximum difference (MD) is found as 15, which defines that the maximum four number of bits have been replaced during the embedding process. The average maximum hiding capacity is found as 1.27 bits-per-pixel for this proposed scheme.

A few attacks have been performed to verify the robustness of the proposed scheme. The extracted watermarks after applying several attacks have been shown in Fig. 7. The quantitative analysis for robustness has been given in Table 2. Results obtained from Fig. 7 and Table 2 together reveal that the proposed scheme is also good in terms of robustness.

The system proficiency is compared to some existing state-of-the-art watermarking methods in Table 3, where, it is observed that the tradeoffs between data capacity and data transparency can be optimized in this proposed scheme.

4   Conclusion

The image watermarking scheme, proposed in this paper, has been performed to embed a color watermark image into a color cover object. An intelligent bit-depth calculation has been deployed along with phase congruency to generate HCM that facilitates adaptive data hiding. This leads to increase in data hiding capacity with high imperceptibility by utilizing the nature of HVS. The qualitative and quantitative results show that this proposed method has successfully overcome the tradeoff between payload and imperceptibility. Moreover, its robustness against several attacks is noteworthy.

References

1. S. Sinha Roy, A. Basu, A. Chattopadhyay, *Intelligent copyright protection for images*, 1st edn. (CRC, Taylor and Francis, New York, USA, 2019)
2. S. Borra, R. Tanki, N. Dey, *Digital image watermarking: theoretical and computational advances*, 1st edn. (CRC, Taylor and Francis, New York, USA, 2018)
3. P.D. Koves, Image features from phase congruency. Videre: J. Comput. Vis. Res. **1**, 1–26 (1999)
4. J.C. Bezdek, R. Ehrlich, W. Full, FCM: the fuzzy c-means clustering algorithm. Comput. Geosci. **10**(2–3), 191–203 (1984)
5. http://sipi.usc.edu/database/database.php?volume=misc
6. M. Kutter, F.A.P. Petitcolas, Fair benchmark for image watermarking systems. in *SPIE, Security and Watermarking of Multimedia Contents*, vol 3657, (CA 1999), pp. 226–239
7. C. Kumar, A.K. Singh, P. Kumar, *Improved wavelet-based image watermarking through SPIHT. Multimedia Tools and Application* (Springer 2018), pp. 1–14
8. S. Han, J. Yang, R. Wang, G. Jia, A robust color image watermarking algorithm against rotation attacks. Optoelectron. Lett. **14**(1), 61–66 (2018)
9. Q. Su, B. Chen, Robust color image watermarking technique in the spatial domain. Soft Comput. Fusion Found. Methodol. Appl. **22**(1), 91–106 (2017)
10. T. Huynh-The, S. Lee, Color image watermarking using selective MSB-LSB embedding and 2D Otsu thresholding. in *International Conference Systems, Man and Cybernetics* (IEEE, Banff Center, Canada, 2017), pp. 1333–1338
11. A. Mishra, A. Goel, A novel HVS based gray scale image watermarking scheme using fast fuzzy-ELM hybrid architecture. in *ELM-2014*, vol. 2 (Singapore, 2015), pp. 145–159

Local Shearlet Energy Gammodian Pattern (LSEGP): A Scale Space Binary Shape Descriptor for Texture Classification

Priya Sen Purkait, Hiranmoy Roy and Debotosh Bhattacharjee

Abstract A novel texture feature is proposed in this paper to classify texture images. To represent the local texture feature in different scales and spaces, a novel local gammodian binary pattern (LGBP) is applied on the shearlet transform domain. The main advantage of the proposed gammodian structure is the size of the feature vector. Again, the LGBP is also very effective in capturing local edge information. Finally, the local shearlet energy gammodian pattern (LSEGP) is proposed. The output result of the proposed LSEGP on Outex database shows the effectiveness of the proposed descriptor in texture classification.

Keywords Shearlet transform · Binary pattern · Texture

1 Introduction

The elementary characteristics of any substance can be easily understood through its texture. Texture gives the wealthiest visual information on the surface of the substance. The repetitive patterns and variations in image intensity levels make texture classification a really challenging problem in image processing. Understanding the intrinsic properties of a substance through its texture is a well-known problem and it is called texture analysis. Texture analysis has many applications such as content-based image retrieval [1], medical image analysis [2], and face analysis [3]. With the

P. S. Purkait · H. Roy (✉)
RCC Institute of Information Technology, Canal South Road, Beliaghata,
Kolkata 700015, India
e-mail: hiranmoy.roy@rcciit.org

P. S. Purkait
e-mail: pspriyapurkait@gmail.com

D. Bhattacharjee
Jadavpur University, Kolkata 700032, India
e-mail: debotosh@ieee.org

increase of volume of visual data, either image or video, the potential application of texture analysis is also increasing rapidly.

Through the years, researchers have developed different methods to extract the texture features through image processing techniques. We can easily categorize those methods into the four following categories:

- **Statistical feature based representation**: In this category, the first- or second-order statistical properties of the distribution of image pixels are considered. Co-occurrence matrix based features were proposed by Hiremath et al. [4] and Ertuzun et al. [5].
- **Transform domain based representation**: In this category, the images are represented in different transformed spaces to hold the texture properties like scale, frequency, and direction. Fourier transform based texture features were represented by Wen et al. [6]. Zhong et al. [7] represented DCT-based texture features. Gabor wavelet based features were proposed by Arivazhagan et al. [8].
- **Model-based representation**: In this category, different stochastic models are used to represent the image textures. Wang et al. [9] proposed a Gaussian and Markov model for representing the texture.
- **Structural approaches based representation**: In this category, the structure of the unit patterns present in the texture is used. A morphological operators based structural representation was proposed by Coco et al. [10]. A comprehensive study on filtering-based structural feature representation can be found in [11].

In recent years, a local binary pattern (LBP) [12] has been gaining huge popularity in texture analysis. LBP is a combination of both statistical as well as structural based feature representation. It captures the local structural patterns by extracting features from a 3×3 neighborhood of a central pixel. Then, it compares the neighbors against the central pixel to capture the statistical relationship between them. In the process of representing the pattern of the local texture, LBP generates a binary string by comparing the central pixel against the neighbors. Excellent results in the field of texture analysis prove the dominance of LBP. However, the normal LBP pattern has its own drawbacks and those drawbacks are (1) sensitive to image noise, (2) only the sign information is considered for representing the string, (3) all patterns are considered, and (4) only circular neighborhood is considered.

Inspired by the output results of LBP and to overcome its drawbacks, many variants of LBP are proposed by many researchers. Tan et al. [13] proposed a local ternary pattern (LTP) to represent more discriminant and noise-resistant texture features. Guo et al. [14] accumulated both the magnitude and sign information and proposed a completed LBP (CLBP). Liao et al. [15] proposed a technique to consider only the patterns, which are the most dominant, i.e., frequently occurring. A Zig-Zag neighborhood based pixel comparison was used in [16] to retrieve the patterns. However, these binary patterns have a very large string (8-bit string).

It has been seen that combining the LBP with other transformations like DCT and wavelet improves the classification result. Therefore, many researchers have proposed different combination features. Chen et al. [17] proposed a combination of DCT and LBP. Tan et al. [18] proposed a combination of Gabor wavelet with LBP. However, DCT and wavelets are not always good at capturing all the inherent patterns.

Shearlet transform [19] has shown its superiority in capturing the inherent geometric patterns better than wavelets and curvelets. The shearlet uses three parameters like scaling, shear, and translation in the affine system to make it suitable to capture the direction of singularities of the texture patterns. LBP-like operators can capture the statistical and structural features of the texture image. If we combine the wavelet-like transformation, which can capture the scale and shape of the texture, along with LBP-like operators, then it becomes a better solution for texture analysis. The different proposed LBP-like representations create a normal feature vector of size 256. Then the addition of wavelet transformation generates a high-dimensional feature. Therefore, a low-dimensional feature is preferred. Motivated by this fact, we proposed a local gammodian binary pattern (LGBP), which generates a very small feature vector of size 16. We applied the LGBP on shearlet transform domain to generate our proposed local shearlet energy gammodian pattern (LSEGP) for texture classification. The proposed descriptor is applied to the Outex database [20] for texture classification; and the output result shows that it is better than other state-of-the-art texture classification methods.

The rest of the paper is organized as follows: in Sect. 2, a detailed description of the proposed method is presented. Experimental results are presented in Sect. 3 and the paper finally concludes in Sect. 4.

2   Proposed Method

In this section, we first explain the idea of digital shearlet transform and its advantages over wavelet and curvelet transform. Then the proposed local gammodian binary pattern (LGBP) and its advantages. Finally, the proposed local shearlet energy gammodian pattern (LSEGP) is explained in detail.

2.1   Shearlet Transform

In order to provide effective instruments for evaluating the inherent geometric characteristics of a signal using anisotropic and directional window functions, the novel directional representation scheme of shearlets appeared [19].

A digital shearlet transform: In digital shearlet transform, $\psi \in L^2(R^2)$ is a set of functions which is defined as

$$\psi_{j,k,m} = 2^{3j/4}\phi(S_k A_{2j} - m) : j \in Z, k \in K \subset Z, m \in Z^2. \tag{1}$$

Here, S_k is the shearing matrix and Z^2 is the digital grid. S_k maps Z^2 onto it to give the treatment of unified continuum and digital setting. K is a thoroughly selected

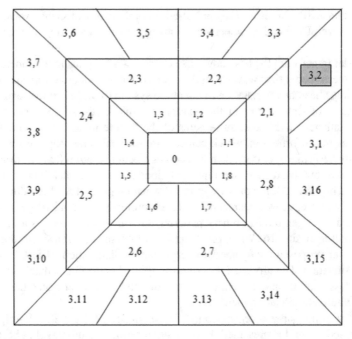

Fig. 1 Four different digital shearlet transformed levels of an image. Here, the zeroth level indicates low-frequency shearlet. Indexes are represented as the level of the transformation and its orientation. For example, index (3, 2) represents the third level of the digital shearlet transform with an orientation value 2

shear indexing set. The digital shearlet system describes a collection of waveforms at different scales (j), controlled orientations (k), and dependent locations (m). A cone-adapted digital shearlet system is defined in Fig. 1. Its cone-adapted form tries to avoid the inherited directionality biased treatment of the system.

$$SH(\phi, \psi, \overline{\psi}) = \phi(. - m) : m \in Z^2 \cup \psi_{j,k,m} : j \geq 0, |k| \leq [2^{j/2}], m \in z^2 \quad (2)$$

where $\overline{\psi}$ is extracted from ψ by swapping both the variables, and $\psi, \overline{\psi}$ and ϕ are functions of L^2.

Advantages over wavelet and curvelet transform: Wavelet depictions are good for point-singular information approximation. When dealing with multivariate data, wavelets are not very efficient. Wavelets fail to handle the discontinuities present in edges of ground borders. Though curvelets can provide better edge response than wavelets, they have some inconveniences. Therefore, some scientists subsequently suggested contourlets, but this method lacks a correct continuum theory.

Shearlet was implemented to solve the constraints of the wavelet by incorporating directional data into it. Three parameters like scaling, shear, and translation have been used in the affine system, which can capture the better direction of edges.

Fig. 2 Gammodian structure and the proposed local gammodian binary pattern (LGBP)

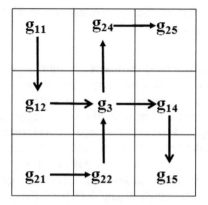

Local Sample Points for LGBP-1 : { g_{11} g_{12} g_3 g_{14} g_{15} }

                                                    8   1   9   7   3

                                LGBP-1 :   0   1   0   0 = 4

Local Sample Points for LGBP-2 : { g_{21} g_{22} g_3 g_{24} g_{25} }

                                                    5   6   9   1   7

                                LGBP-2 :   1   1   0   1 = 13

2.2 Local Gammodian Binary Pattern

Local Binary Pattern (LBP) is a straightforward and highly effective texture operator that labels picture pixels by thresholding each pixel's neighborhood and sees the outcome as a binary number. One of the major problems of LBP is the length of the feature vector (256). Again, normal LBP is not suitable to measure the smaller segments or lines present in the texture pattern. Therefore, we proposed a new binary pattern, which considers the image pixels present in a 3×3 window in the form of a gammodian structure (as shown in Fig. 2). This gammodian structure is able to capture the line segments present in the texture. At the same time it has a very small size of feature vector 16.

There are two patterns consisting of two different pixel sequences (as shown in Fig. 2). One pattern (LGBP-1) consists of g_{11}, g_{12}, g_{13}, g_{14}, g_{15} pixels and another of g_{21}, g_{22}, g_{13}, g_{24}, g_{25}. The binary pattern is generated using the following equation:

$$LGBP_k = \sum_{i=1}^{4} f(g_{k,i+1} - g_{k,i}) \times 2^{i-1}, k \epsilon \{1, 2\} \tag{3}$$

$$f(x) = \begin{cases} 1, x >= 0 \\ 0, otherwise. \end{cases}$$

2.3 Local Shearlet Energy Gammodian Pattern Representation

Shearlet transform captures the inherent texture information in different scales and spaces at coarse level. Then, we apply LGBP in the transform domain to capture the statistical and structural pattern of the shearlet domain information. Since the shearlet transform gives output in the form of coefficients (values ranging from negative to positive), we apply a local energy computation of them. Again to reduce the feature vector, we select the best shearlet scale space features, which have high entropy. Here, we select the best six scale space information among the 16. Then, we apply LGBP for them to get our proposed local shearlet energy gammodian pattern (LSEGP). Figure 3 shows the proposed system in pictorial form.

2.4 Similarity Measure

The similarity between two different texture images is measured in the LSEGP domain using a simple NNC classifier with χ^2-distance similarity measure. Here we have used the 16 binary bins for two different LGBP patterns and six different shearlet transform domains. Therefore, the final length of the LSEGP descriptor is $2 \times 16 \times 6 = 192$.

3 Results and Discussion

Here we use the most common Outex database [20] to evaluate the performance of the proposed LSEGP. This database has images of different surface textures with normal conditions (Outex_TC10), varying illuminations (Outex_TC12_000), and rotations (Outex_TC12_001). All the three benchmark data sets are used in the experiment of texture classification. The performance of the proposed LSEGP is compared with other existing state-of-the-art texture classification methods: LBP [12], LTP [13], CLBP [14], DLBP [15], and LDZP [16]. The proposed LSEGP provides better results (Table 1). One sample texture image with different stages of the proposed LSEGP representation is shown in Fig. 4.

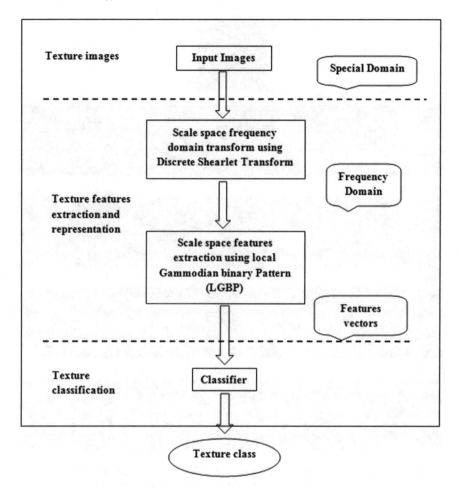

Fig. 3 Overview of the proposed texture classification system

4 Conclusion

A novel local gammodian binary pattern is proposed to represent the patterns present in the texture. The shearlet transform is used to capture different scale space features. The proposed LSEGP gives superior results on the texture database. In future, LSEGP can be tested with noisy images and other applications related to image processing.

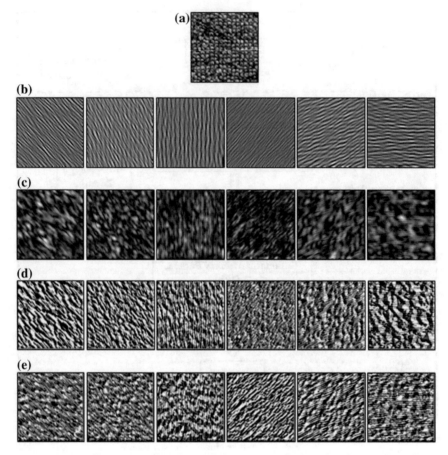

Fig. 4 One sample image from Outex database with different LSEGP patterns. **a** Original Outex database image, **b** Six selected shearlet transform images, **c** Local energy calculation for the selected shearlet transform images, **d** LGBP-1 images, and **e** LGBP-2 images

Table 1 Classification results(%) on Outex database

Methods	TC10	TC12_000	TC12_001
LBP	84.89	63.75	65.30
LTP	76.06	63.42	62.56
CLBP	94.53	82.52	81.87
DLBP	98.10	87.40	91.60
LDZP	99.95	99.93	99.82
Proposed LSEGP	**99.98**	**99.97**	**99.97**

References

1. B.S. Manjunath, W.Y. Ma, Texture features for browsing and retrieval of image data. IEEE Trans. Pattern Anal. Mach. Intell. **18**(8), 837–842 (1996)
2. M. Peikari, M.J. Gangeh, J. Zubovits, G. Clarke, A.L. Martel, Triaging diagnostically relevant regions from pathology whole slides of breast cancer: a texture based approach. IEEE Trans. Med. Imaging **35**(1), 307–315 (2016)
3. T. Ahonen, A. Hadid, M. Pietikainen, Face description with local binary patterns: application to face recognition. IEEE Trans. Pattern Anal. Mach. Intell. **28**(12), 2037–2041 (2006)
4. P.S. Hiremath, S. Shivashankar, Wavelet based co-occurrence histogram features for texture classification with an application to script identification in a document image. Pattern Recognit. Lett. **29**(9), 1182–1189 (2008)
5. A. Latif-Amet, A. Ertuzun, A. Ercil, An efficient method for texture defect detection: sub-band domain co-occurrence matrices. Image Vis. Comput. **18**, 543–553 (2000)
6. C.Y. Wen, R. Acharya, Self-similar texture characterization using a Fourier-domain maximum likelihood estimation method. Pattern Recognit. Lett. **19**(8), 735–739 (1998)
7. D. Zhong, I. Defee, DCT histogram optimization for image database retrieval. Pattern Recognit. Lett. **26**(14), 2272–2281 (2005)
8. S. Arivazhagan, L. Ganesan, S.P. Priyal, Texture classification using Gabor wavelets based rotation invariant features. Pattern Recognit. Lett. **27**(16), 1976–1982 (2006)
9. L. Wang, J. Liu, Texture classification using multi-resolution Markov random field models. Pattern Recognit. Lett. **20**(2), 172–182 (1999)
10. K.F. Coco, E.O.T. Salles, M.S. Filho, Topographic independent component analysis based on fractal and morphology applied to texture segmentation, in Proceedings of the international conference on independent component analysis and signal separation (ICA 2009), pp. 491–498 (2009)
11. T. Randen, J.H. Husoy, Filtering for texture classification: a comparative study. IEEE Trans. Pattern Anal. Mach. Intell. **21**(4), 291–310 (1999)
12. T. Ojala, M. Pietikainen, T. Maenpaa, Multiresolution gray-scale and rotation invariant texture classification with local binary patterns. IEEE Trans. Pattern Anal. Mach. Intell. **24**(7), 971–987 (2002)
13. X. Tan, B. Triggs, Enhanced local texture feature sets for face recognition under difficult lighting conditions. IEEE Trans. Image Process. **19**(6), 1635–1650 (2010)
14. Z. Guo, L. Zhang, D. Zhang, A completed modeling of local binary pattern operator for texture classification. IEEE Trans. Image Process. **19**(6), 1657–1663 (2010)
15. S. Liao, M.W. Law, A.C. Chung, Dominant local binary patterns for texture classification. IEEE Trans. Image Process. **18**(5), 1107–1118 (2009)
16. S.K. Roy, B. Chanda, B.B. Chaudhuri, S. Baneerjee, D.K. Ghosh, S.R. Dubeye, Local directional ZigZag pattern: a rotation invariant descriptor for texture classification. Pattern Recognit. Lett. **108**, 23–30 (2018)
17. P. Chen, S. Chen, A new face recognition algorithm based on DCT and LBP, in book: ed. by B. Cao, G. Wang, S. Chen, S. Guo. Quantitative Logic and Soft Computing, Advances in Intelligent and Soft Computing, vol. 82, (Springer, Berlin, Heidelberg, 2010)
18. X. Tan, B. Triggs, Fusing gabor and LBP feature sets for Kernel-based face recognition, in *Proceedings of the International Workshop on Analysis and Modeling of Faces and Gestures (AMFG 2007)*, pp. 235–249 (2007)
19. G. Kutyniok, W.Q. Lim, and X. Zhuang, Digital shearlet transforms, in *Shearlets: Multiscale Analysis for Multivariate Data* (Springer, Berlin 2012)
20. T. Ojala, T. Maenpaa, M. Pietikainen, J. Viertola, S. Huovinen, Outex-New framework for empirical evaluation of texture analysis algorithms, in *Proceedings International Conference on Pattern Recognition (ICPR 2002)*, pp. 701–706 (2002)

Printed in the United States
By Bookmasters